4th Nationwide Scientific Conference on Engineering-Infrastructure-Mining (BIG 2018)

E3S Web of Conferences Volume 36 (2018)

Krakow, Poland
11-12 January 2018

Editors:

E. Pilecka
J. P. Kogut
T. Tatara

ISBN: 978-1-5108-6102-2

Printed from e-media with permission by:

Curran Associates, Inc.
57 Morehouse Lane
Red Hook, NY 12571

Some format issues inherent in the e-media version may also appear in this print version.

Printed by Curran Associates, Inc. (2018)

For additional information, please contact EDP Sciences – Web of Conferences at the address below.

EDP Sciences – Web of Conferences
17, Avenue du Hoggar
Parc d'Activité de Courtabœuf
BP 112
F-91944 Les Ulis Cedex A
France

Phone: +33 (0) 1 69 18 75 75
Fax: +33 (0) 1 69 28 84 91

contact-edps@webofconferences.org

Additional copies of this publication are available from:

Curran Associates, Inc.
57 Morehouse Lane
Red Hook, NY 12571 USA
Phone: 845-758-0400
Fax: 845-758-2633
Email: curran@proceedings.com
Web: www.proceedings.com

TABLE OF CONTENTS

Statement of Peer review

In submitting conference proceedings to *Web of Conferences*, the editors of the proceedings certify to the Publisher that

1. They adhere to its **Policy on Publishing Integrity** in order to safeguard good scientific practice in publishing.
2. All articles have been subjected to peer review administered by the proceedings editors.
3. Reviews have been conducted by expert referees, who have been requested to provide unbiased and constructive comments aimed, whenever possible, at improving the work.
4. Proceedings editors have taken all reasonable steps to ensure the quality of the materials they publish and their decision to accept or reject a paper for publication has been based only on the merits of the work and the relevance to the journal.

Title, date and place of the conference

BIG 2018 - 4th Nationwide Scientific Conference on Engineering-Infrastructure-Mining

January 11-12, 2018, Kraków, Poland

Proceedings editor(s):

Elżbieta Pilecka, Tadeusz Tatara, Janusz P. Kogut

Date and editor's signature

28-03-2018

E3S Web of Conferences **36**, 01001 (2018) https://doi.org/10.1051/e3sconf/20183601001
BIG 2018

Improving the design of the earthquake-proof suspension building

Taliat Azizov[1], *Nadzieja* Jurkowska[2]

[1]Pavlo Tychyna Uman State Pedagogical University, Uman, Ukraine

[2] Tadeusz Kosciuszko Cracow University of Technology, Krakow, Poland

Abstract. The proposal for improving the constructive scheme of the supporting frame of the suspension building has been considered. The strains in its bearing elements due to earthquakes are significantly less than in the traditional cantilever buildings. In former schemes the vertical components of strains develop during horizontal oscillations of the grounds. This disadvantage has been eliminated now. It is shown that in the improved design of the structure under the earthquake the vertical component of vibrations is practically absent, due to arrangement of the suspension device strictly vertically. The numerical data for calculation results with variation of some characteristics of the supporting frame and suspensions have been shown in order to allow adjusting the dynamic strains at the vibrations of the base. The scheme for installation the vertical elements of the frame have been proposed. Its usage will allow to build the high-rise suspension buildings.

Keywords: suspension building, seismic safety, dynamic forces, Lagrange's equations of the second kind

1 The problem statement

The issue of protecting buildings subjected to earthquakes is always one of the most important tasks for design and construction in the seismic areas. The usage of the damping devices, disconnecting constraints is an effective protection measures [1-5]. However the dynamic forces from the action of the earthquake still large enough, so that strengthening of the building structures is required, and the cost of construction is increasing.

Another highly effective way for the substantial reduction of the dynamic forces acting on the bearing structures is the suspension of the building to the bearing frame [6,7]. According to the preliminary calculations by the simplified dynamic schemes, the suspension of the very building allows to reduce the loading of the bearing structures several times. The unusual solution to the problem of the suspension building entails facilitation in installation of structure of the bearing frame, improving its shape to reduce the vertical components of force during oscillation, and a study of change of the efforts to the support frame depending on various factors, including the length of the suspension thread and the stiffness of the supporting frame.

Thus, the purpose of this article is to improve the structure of the supporting frame of a suspension building. The tasks presuppose development of a technique for its simple installation without usage of special tools and the analysis of the influence of some construction parameters on dynamic efforts.

2 Improving the supporting frame of the suspension building

The building proposed in [6] is a structure suspended on the bearing frame, which mass is much less than the mass of the very building. The preliminary calculations are conducted by the building mechanics' techniques using the Lagrange equations of the second kind. Herewith two options for the suspension of the building have been proposed: the pyramidal and the rectangular frames are used as the bearing structures (Figure 1). In both cases, the building itself is suspended to the support frame on the cables.

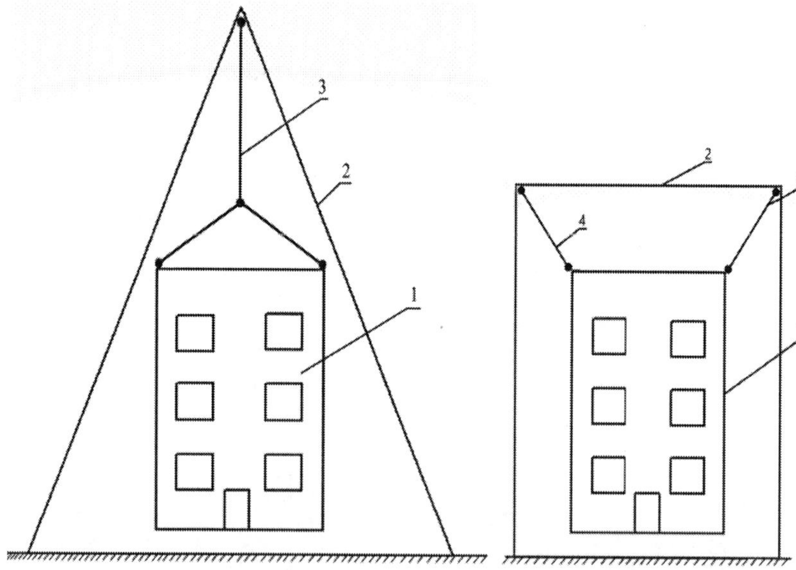

Fig. 1. The scheme of the suspension building:
a) suspension on the pyramid frame; b) suspension on the rectangular frame
1 – suspension building; 2 – frame; 3, 4 – suspension (cables)

The advantage of the rectangular frame is that its area is much less than the area of the pyramidal frame. However the large vertical components of the strains develop in the cables during the horizontal oscillation, due to the inclined arrangement of the cables, on which the building is suspended.

The principle of the preliminary dynamic calculation for such a suspension building has been explained in [7]. Its purpose is to determine the equations of motion of the system and of the order of the seismic forces acting on the bearing frame. In this case, the whole suspension building (or structure) is considered as a point mass, suspended on the bearing frame. A simplified, but the most approximate dynamic calculation scheme for this case is presented in Figure 2.

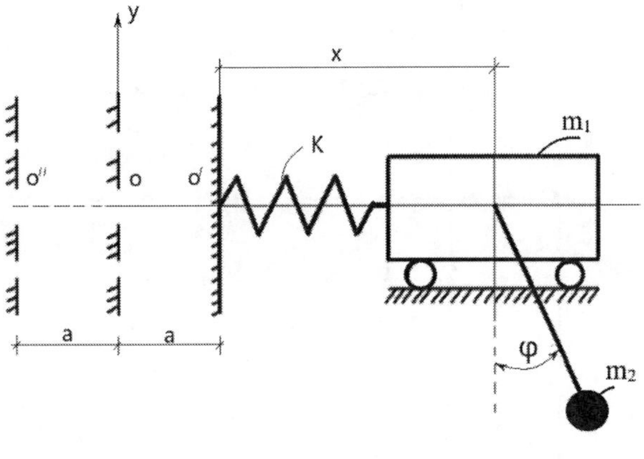

Fig. 2. Dynamic scheme with the oscillating support

The support located at the beginning of the coordinate system OXY, oscillates between the points O^I and O^{II} relatively to the equilibrium position O by the law $x = A\sin(pt)$, where a is the amplitude, p is the frequency. Amplitude and frequency of oscillations can be determined by seismograms, depending on the seismic zone.

The cart is attached to a support by a spring of stiffness k. The conditional stiffness of the spring is determined from the condition $x_c=x_s$ (Figure 3).

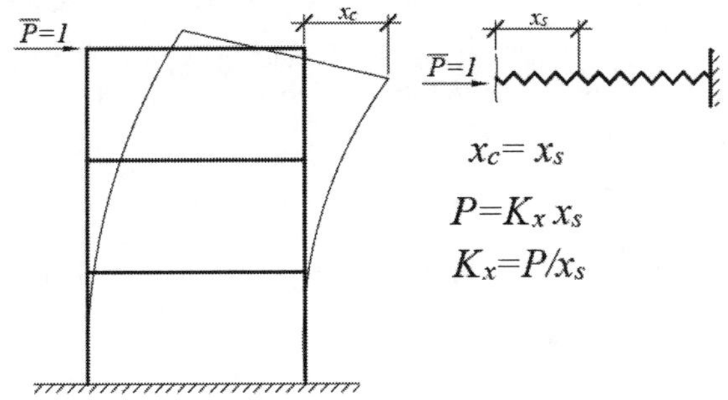

Fig. 3. Dynamic scheme with the oscillating support

The spring stiffness simulates the displacement of the free end of the frame in horizontal direction and is chosen from the condition of equality of the static deflection of the end of the cantilever (supporting frame) from a single horizontal force to the displacement of the spring from the same force. The base of the system (point O in Figure 2) oscillates horizontally left and right with an amplitude $a=O\text{-}O^I$.

It should be noted that this scheme is valid for the pyramidal frame (construction according to Figure 1, a), but is not quite applicable for the design of the rectangular frame (Figure 1, b). The reason is that when using the scheme according to Figure 1, b, at the horizontal oscillation of the foundation considerable vertical stress components develop in the suspensions. They should be considered at the design. At the same time the pyramidal frame takes much place.

In order to eliminate this disadvantage, the structure of the bearing frame is proposed to be changed (Figure 4). The difference from the Figure 1,b is in the vertical placement of the suspensions 3 and 4, and in the arrangement of the additional bracing in the bearing frame. Therefore, during the horizontal oscillations the foundation will also oscillate in the horizontal direction, and the vertical component of dynamic stress will be practically absent.

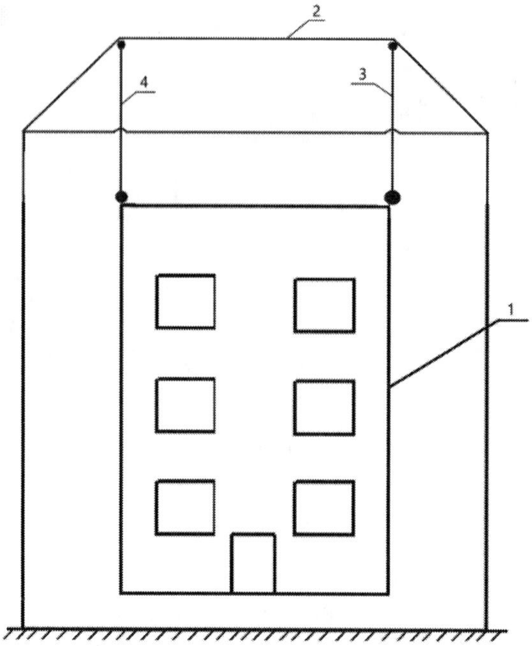

Fig. 4. The improved scheme of the bearing frame of the suspended building
1 – suspended building; 2 – the bearing frame; 3, 4 – suspension (cables)

Since according to the scheme on the Figure 4 for the small oscillations of the foundation the vertical stresses will be absent in the suspensions of the system, the usage of the dynamic scheme represented on the Figure 2 will be effective.

This scheme has been calculated using the Lagrange equations of the second kind [7]. The derived system of equations is:

$$\begin{cases} (m_1 + m_2)\ddot{x} + m_2 l\ddot{\varphi}\cos\varphi - (m_1 + m_2)p^2 a \sin pt + kx = 0 \\ \ddot{x}\cos\varphi + l\ddot{\varphi} - ap^2 \sin pt \cos\varphi + g\sin\varphi = 0 \end{cases} \quad (1)$$

where, according to the Figure 2: m_1, m_2 are the masses of the trolley (simulating the bearing frame) and the suspension of the building, respectively; x and φ are the generalized coordinates; a and p are the amplitude and frequency of oscillation of the base, respectively; g is the acceleration of gravity.

The system of differential Lagrange's equations can be easily solved both numerically and analytically. Knowledge of the law of motion of this system allows to obtain the forces acting on the system's elements, including the spring (the horizontal component to the supporting frame in this case).

3 The technique for installation the construction of the supporting frame

The length of the columns of the bearing frame when suspending the high building is large enough. The techniques for installation the high-rise buildings by the increasing of separate sections or rising the building as a whole are well known. During the lifting entirely, the falling masts are used. Their height is accepted equal to about a third of the height of mounted support. In this regard the complexity of works and their cost increase. In case of installation of the supporting frame of the suspension building, it is possible to avoid the falling mast, simplifying the installation essentially. The arrangement scheme is shown in Figure 5.

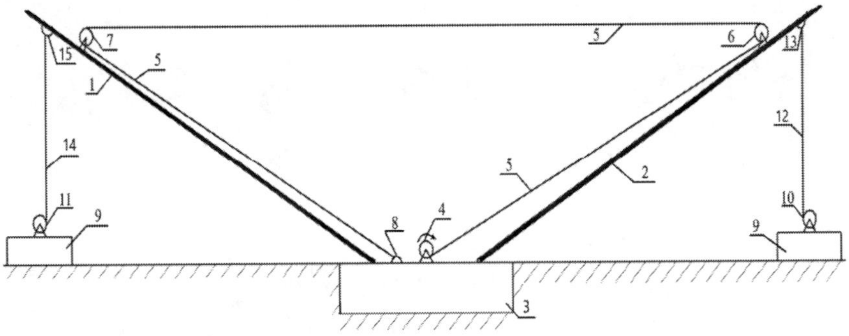

Fig. 5. The arrangement of installation of the columns of the supporting frame of the suspension building

The pillars left 1 and right 2 of the paired columns of the bearing frame are pivotally mounted to the monolithic foundation 3 (in the direction perpendicular to the drawing, the columns are coupled, i.e. two columns rise right, and two columns rise left). The free ends of the columns 1 and 2 should be lifted beforehand to the heights h and placed on temporary supports. The winch 4 is installed on the foundation. The cable of the winch 5 is held through the block 6 fixed on the right column, and through the block 7 fixed on the left column, and is attached to the foundation at the point 8. Thus, it is possible to lift simultaneously both of the coupled columns.

In order to eliminate the uneven lifting, the temporary counterweights 9 are mounted under the columns. The winches 10 (at the right column) and 11 (at the left column) are fixed on these columns. The cable 12 from the winch 10 is mounted at the point 13 of the column on the right. Similarly, the cable 14 from the winch 11 is attached at point 15 on the left column.

When the leading winch 4 is enabled, in case the balance is disturbed and one of the columns rises more than the other, the balance should be recovered by means of wires 12 and 14. During the installation process the cables emerging from winches 10 and 11 should be tightened. Such a scheme allows to lift both of coupled columns simultaneously, as well as prevents rollover of the system.

The arrangement scheme (see Figure 5) is very effective. The presence of the unit 6, through which the lifting cable 5 is mounted to the opposite column, sharply reduces the stress required for lifting of both columns. It can be easily seen from the design diagram, presented in Figure 6 (the right column is presented, the left column will be symmetrical).

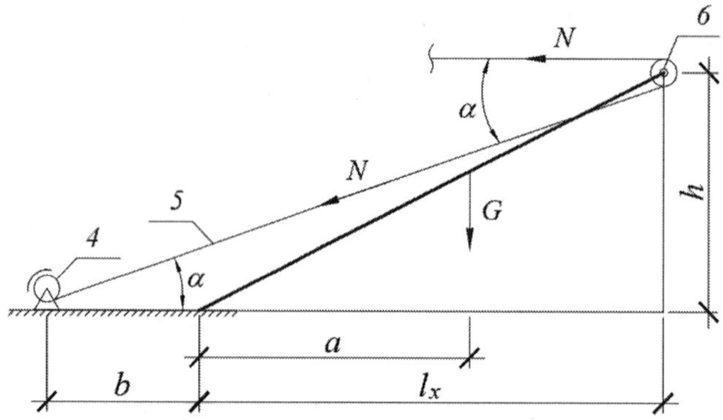

Fig. 6. The design scheme for lifting the columns to the design position

Considering the power winch (p.4 in Figure 6), the strains in the cables can be calculated from equilibrium conditions. There are two options: 1. for the cable attached at point 6 of the right support and not thrown to the left support (see also Figure 5); 2. by the proposed scheme using units 6 and 7, and with the cable attached at point 8.

In order to fix the elevating part of each column in vertical position, the cantilever elements must be welded to them. Those cantilever elements must be perpendicular to the elevating columns' parts and of the length equal to half the distance between columns. After bringing columns to the vertical position the cantilever elements of two columns are fastened together by bolts or welding.

The calculations using equation (1) demonstrate several times decreasing of the strains in the bearing structures of the suspension building, comparatively to the strains in the building constructed by the traditional cantilever scheme.

4 The impact of individual factors on the stresses in the system

The calculation has been conducted for the scheme of the suspended building represented in Figure 2, on the basis of solution for the derived system of Lagrange's equations of the second kind (1). Initial data: the mass of the trolley (i.e. the mass of the bearing frame) m1=51000 kg, mass of the building m2=1612000 kg. These data were obtained by a preliminary rough calculation of the five-storey building with dimensions 12×12 m and a total mass 750 kg/m2 (including the mass of structures, columns, floors, etc.). The columns of the supporting frame are accepted from steel hollow tubes with a thickness of 10 mm. The basic value of the stiffness coefficient of the conditional spring k (the stiffness of the supporting frame in the horizontal direction) was determined from the calculation of the bearing frame on a horizontal unit load, and is equal to kb=2303000 N/m. Let the oscillation amplitude of the ground a=0.048 m, the oscillation frequency p=10 Hz; the maximum acceleration of the ground vibration w=4.8 m/s2 maximum velocity of ground vibration vmax=0.48 m/s (approximate data for the 9 magnitude earthquake by the Richter scale).

For the suspension thread of length 5 m the change of the generalized coordinates $x=x(t)$ and $\varphi=\varphi(t)$ for the first five seconds of oscillations is shown in Figure 7.

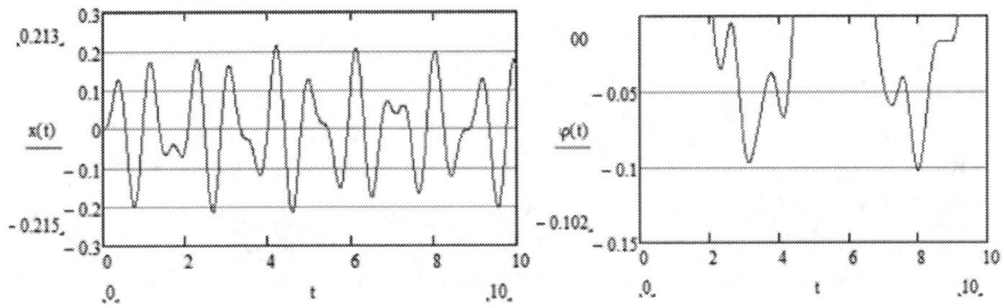

Fig. 7. Graphs of change
a) of the variable $x=x(t)$ of the displacement of the point of the frame's suspension
b) of the rotation angle $\varphi=\varphi(t)$ of the thread suspension

The rotation angle is sufficiently small, i.e. even in case of strong earthquake deviation of the buildings from the vertical position does not exceed 4°. Herewith the horizontal force is small enough.

From the solution of the system of differential equations (1), the oscillation dependences $x=x(t)$ of the conditional truck (Figure 2) have been obtained. The maximum values xmax are taken from these dependences. The maximum horizontal force on the bearing frame is $F_{max} = x_{max} \cdot k$. Meanwhile, for the cantilever (traditional) building with acceleration of oscillation of ground w=4.8 m/s2 the total horizontal force will be equal to $F_{tr} = w \cdot m_2 = 7737600\ N$.

The table 1 provides the calculated data for different length of the suspension of building l (the length of the suspension cables) and different stiffness of the supporting frame k, which is accepted for both less and greater than the value basic k_b (see above). The values of the ratio of maximum horizontal force of the traditional cantilever building F_{tr} to the horizontal force of the suspension building F_{max} are presented in the last column.

Table 1. Ratio of maximum horizontal forces of the cantilever and suspension buildings.

№	l, m	k, N/m	x_{max}, m	F_{tr}, N	F_{max}, N	F_{tr}/F_{max}
1	3	$k=k_b$=2303000	0.235	7737600	541205	14.29
2	5	$k=k_b$	0.213	7737600	490539	15.77
3	7	$k=k_b$	0.213	7737600	490539	15.77
4	10	$k=k_b$	0.198	7737600	455994	16.97
5	12	$k=k_b$	0.201	7737600	462903	16.71
6	15	$k=k_b$	0.204	7737600	469812	16.46
7	5	$k=k_b*0.6$	0.178	7737600	245960	31.45
8	5	$k=k_b*0.7$	0.202	7737600	325644	23.76
9	5	$k=k_b*0.8$	0.187	7737600	344529	22.46
10	5	$k=k_b*0.9$	0.202	7737600	418685	18.48
11	5	$k=k_b*1.1$	0.252	7737600	638391	12.12
12	5	$k=k_b*1.2$	0.273	7737600	754463	10.25
13	5	$k=k_b*1.3$	0.231	7737600	691591	11.19
14	5	$k=k_b*1.4$	0.241	7737600	777032	9.95
15	5	$k=k_b*1.5$	0.265	7737600	915442	8.45

The table demonstrates the possibility of reducing the horizontal forces by reducing the stiffness of the bearing frame. Herewith the frequency and period of oscillation of the system also change.

In order to exclude resonance phenomena during an earthquake, the strains in the bearing frame, as well as the frequency and period of oscillation of such a building can be regulated by changing the length of thread of the suspension, stiffness of the bearing frame in the horizontal direction,2 ratio of the mass of the frame and the period of oscillation of such a building.

Calculations illustrate that the horizontal component of the dynamic force on the bearing frame of the suspension building during horizontal oscillation of the ground is significantly less than the similar strength to the traditional cantilever building. Moreover, unlike the protection systems with disconnected constrains, the proposed system does not require replacement of any structural elements after the earthquake.

One of essential advantages of these proposals is the possibility to conduct numerous preliminary calculations by the approached schemes. Thus the most acceptable design option for the structure of the bearing frame, suspensions and other parameters can be evaluated. After the preliminary determination of the structural scheme, the final calculation of strength, stability and deformability can be conducted using the known software systems such as Ansys, Abacus etc.

5 Conclusions and perspectives of research

The mass of the proposed suspension building is many times greater than the mass of the supporting frame. This factor together with the design features of the proposed kinematic scheme of the building allow to reduce dynamic forces in bearing structures of the building during an earthquake significantly. The article demonstrates that the change in stiffness of the supporting frame in the horizontal direction results in regulation of strains in the bearing structures, the frequency and period of oscillation of the building during an earthquake. Additionally, the proposed technique of installation of the supporting frame allows to construct buildings of such design, including high-rise buildings without the use of sophisticated equipment, helicopters, etc. In totality, this makes the proposed building attractive both from a perspective of seismic safety, and from the position of relative simplicity of construction.

In future it is planned to continue improvement of techniques of dynamic calculation of the suspension buildings, analysis of impact of various factors on the strains, deformability, the frequency of oscillation of the building during the earthquake, as well as the development of various details of the proposed structural system.

References

1 C.C. Chang, Mass dampers and their optimal designs for building vibration control, Eng. Struct. 21(5) (1999)

2 Y. Eisenberg, Constructions with disconnective constraints for seismic areas [Sooruzheniya s vyklyuchayushchimisya svyazyami dlya sejsmicheskih rajonov]. Moscow (1976).

3 R.O. Hamburger, Facts for Steel Buildings Number 3 – Earthquakes and Seismic Design, 71 (2009).

4 K.S. Moon Structural Design of Double Skin Facades as Damping Devices for Tall Buildings, Procedia Eng. (2011) 1351–1358.

5 Eurocode 8: Seismic Design of Buildings Worked examples, (2012).

6 T. Azizov, Design of the earthquake-resistant buildings [Konstrukciya sejsmichno stijkoï budivli]. Patent for the useful model Nr 54247. Ukraine, Bulletin Nr 20.

7 T. Azizov, Dynamic design scheme of suspended seismically safe buildings, in: Sci. Eur. 17, Vol 1 (2017)

Modeling of information on the impact of mining exploitation on bridge objects in BIM

Piotr Bętkowski[1,*],

[1]Silesian University of Technology, Faculty of Civil Engineering, Department of Mechanics and Bridges, Gliwice, Poland

Abstract. The article discusses the advantages of BIM (Building Information Modeling) technology in the management of bridge infrastructure on mining areas. The article shows the problems with information flow in the case of bridge objects located on mining areas and the advantages of proper information management, e.g. the possibility of automatic monitoring of structures, improvement of safety, optimization of maintenance activities, cost reduction of damage removal and preventive actions, improvement of atmosphere for mining exploitation, improvement of the relationship between the manager of the bridge and the mine. Traditional model of managing bridge objects on mining areas has many disadvantages, which are discussed in this article. These disadvantages include among others: duplication of information about the object, lack of correlation in investments due to lack of information flow between bridge manager and mine, limited assessment possibilities of damage propagation on technical condition and construction resistance to mining influences.

Keywords: BIM, mining damages, bridges, management, maintenance

1 Introduction

Traditional model of managing bridge infrastructure on mining areas has many disadvantages, which are discussed below in a few next points. Basic problem is the lack of proper connection of many information related to the bridge objects and the lack of information flow between entities interested in maintaining the bridge properly (e.g. road manager, local mine, commune/city office, technical infrastructure administrators).

In Poland and in many other countries methods of modeling information about building objects, know under common password BIM (Building Information Modeling) are widely popular [e.g. 1÷5]. The application of BIM or the implementation of some of the assumptions are mandatory (i.e. required by law) e.g. in Great Britain, USA, Norway, Denmark, Singapore. In Poland several BIM pilot projects in the field of design, construction and maintenance of bridges are implemented and are being implemented.

* Corresponding author: piotr.betkowski@polsl.pl

In the next few years, BIM will be gradually introduced in all countries associated in the European Union, in accordance with the requirements of the European Directive 2014/24/EU announced in February 2014 on competitive tendering. In the first place, BIM will include public investments, and thus bridge objects. Regardless of law regulations imposed, BIM is a very efficient method of information modeling, allowing optimizing management activities throughout the entire life cycle of bridges. BIM technology can be implemented for mining damages, using available methods, algorithms and software. The article discusses the advantages of BIM technology in relation to bridge objects located on mining areas. It is shown how the correct modeling of information about the object helps to reduce mine costs related to the removal of mine damages and protection of objects against mining influences and reduces the costs of administration relate to the maintenance of the objects. Other advantages of having a common advanced BIM model of the bridge are also presented, such as the possibility of remote monitoring, improvement of safety, coordination of mining operations and managing of the bridge objects in the field of renovation and modernization, improvement of the relationships among the bridge manager, coal mine and local commune/city office.

2 Basic problems related to bridges on mining areas

Administrators of roads generally do not have full information about the impact of mining deformations on bridge objects over time, e.g. deformation indicators (horizontal strains ε, etc.) for every months or the continuous function of settlements w, but only the end values of indicators at the end of a certain period of time (e.g. the Three Years' Mine Exploitation Plan). The inability controlling the growth of deformation causes distrust to mines and makes difficult the proper supervision of the object, e.g. during the appearance of mining influences on the surface of the area.

Mines have a little knowledge about the construction of bridges and investment plans of Road Administration. In areas of mining influences, certain information should be available for mines; such data could be helpful in planning coal mining exploitation in a given mining wall over time, in activities related to removal of mining damages and mining prevention. Due to the improper information flow, the mines many times prepared the documentation of removal mining damages and adaptation of the structures to taking over future mining influences for objects planned to replacement by the Administrator due to insufficient load capacity or insufficient gauge. In many cases, the object from an economic point of view could be protected for a shorter period of time with only significant damages removed, if, for example, the rebuilding is planned by the Administrator in three years.

The static and dynamic resistance of the bridge objects to mining influences depend on its technical conditions and kinematic capabilities. Mines make the assessment of the resistance of objects to mining influences once every three years as part of so-called Mine Exploitation Plan. During that time, several mining walls may be exploited in the area of the bridge, and damages may occur in the object - that reduces resistance to mining influences, e.g. significant cracks in the zone of joining the pillars and spans in the frame constructions [6, 7]. The static and dynamic resistance of bridge structures to mining influences is not constant, but changes in time. The assessment of the technical condition is, in the case of mining areas, also the assessment of damage propagation and the assessment of the impact of damages on the "work" of the structure. Using BIM, damages can be easily taken into account in the bridge model and updated if necessary.

Using tools and technologies that are easily combined with BIM, such as e.g. AR (Augmented Reality), we can overlap the images from the last and current bridge review (tablet, projection glasses) during visit on the object, which allows quickly and reliably control critical sections directly on the real construction (e.g. [4, 5]).

The article shows the problems with information flow in the case of bridge objects located on mining areas and indicates the benefits of advanced information models (BIM), e.g. the possibility of automatic monitoring of structures, improvement of safety, optimization of maintenance activities (e.g. repairs) by time and scope (cost reduction). Bridge objects management can be understood as combination of all technical, administrative and management activities of the life cycle after the completion of the construction of the object.

3 BIM as information model

At present, flat 2D drawings dominate in relation to bridge infrastructure in mining areas (e.g. [6 ÷ 13]) and popular are mathematical formulas given by professor Rosikoń [14]. More possibilities are given by use of 3D spatial model, extended by additional parameters, so that the rotations and displacements of spans and bridgeheads can be modeled as rigid solids [15]. Spatial models of the objects and deformations allow directly on include information from monitoring systems into the bridge models and using measurements from various sources and methods. Efficient tools are provided by BIM technology [1 ÷ 5]. By supplementing the 3D BIM model with additional parameters, we can quickly obtain the 4D model (work schedule) and 5D (investment cost). In this way it is easy to analyze and compare cost-effectiveness and organisation of several repair solutions (currently technical-economic comparisons are rarely carried out). BIM 6D enables quick realization of accurate energy analyzes for buildings, but bridges do not required such analyzes. BIM 7D is used by managers in the process of managing the building objects at the stage of its exploitation. BIM tools allow on describing in a certain period of time/future, the investment plans of the bridge objects manager and planes activities of the mine in the scope of damage repair and prevention, in such case the correlation of investments is easier.

BIM technology is a process, during which a model common for all industries is created. This model combines additional information related to the individual elements of bridge (e.g. expansion joints, bridge bearings). In such process, a number of additional parameters may be included such as, for example, results of geodetic measurements, data from automate measurement (monitoring) systems and other technical information needed by each of the participants (e.g. investment plans of the bridge administration, activities planned by the local mine). All information that are included in this model can be used by all participants, such as the bridge administration, local commune or city, local mine, technical infrastructure administrators running after (under) the bridge (e.g. waterworks, gas pipeline, power cables) or an institution responsible for water management in the case of watercourses. When talking about the process, we should keep in mind the organization of flow of information in the amount needed by each participant.

The traditional model of managing bridge infrastructure on mining areas is inefficient, it is necessary to contact each participant with each other, in order to obtain information (Fig. 1.). Moreover each participant may have different information, often out of date. These disadvantages cause distrusts between the bridge administration, the mines and the local self-government units. These disadvantages make also difficult planning of exploitation of coal deposits in the area of the bridge, negatively affect on safety and comfort of use, and increase the cost of maintaining objects and the removal of damages.

BIM is a technology based on a common model to which all participants of the process can access in real time (Fig. 2.). The model works properly if all participants update the data which they have provided to the model. Participants can expand the model with additional data at any time. Interoperability of participant's technology platforms is important here, i.e. providing data in a form that all participants can use. It is necessary to standardize the procedures and details of the model, e.g. marking and description of layers, because all participants of the process have access to the same model. Ready solutions in the field of cooperation (interoperability) are provided by, for example, the British family of BS 1192 standards, defining the framework of BIM processes.

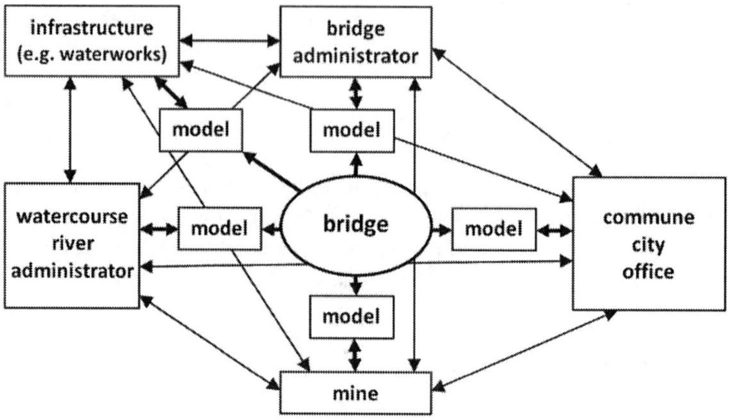

Fig. 1. Traditional model of information flow (bridge object on the mining area).

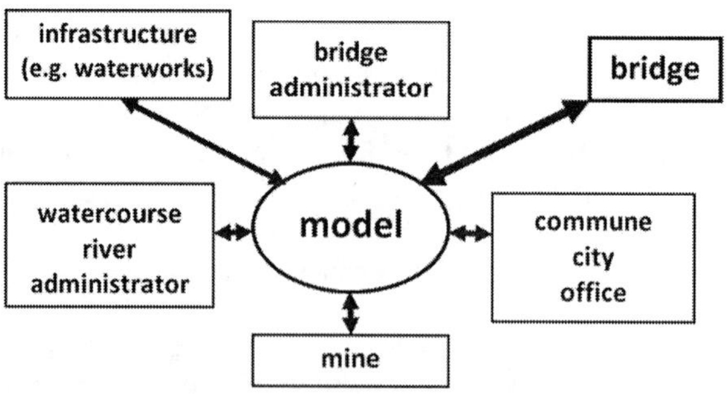

Fig. 2. BIM model of information flow (bridge object on the mining area).

4 Duplication of information about the bridge object

Every three years, in the scope of Mine Exploitation Plan, each of the mines performs documentation containing the assessment of static and dynamic resistance to mining influences of all bridge structures located in its mining area, which are located in the reach of the planned mining exploitation. The contractor for such elaboration is selected in the open tender. Practically every time, the bridge geometry and the damages of structure are inventoried. In the case of BIM technology, the model of the bridge with current structural damages would be available for downloading, so the workload for such assessment would drop by more than 50% (the major part of the documentation are the same information obtained all over again by another company). In addition, it would be possible to analyze the propagation of damage over time (e.g. indication of active cracks, etc.), so the quality of the documentation would be significantly improved.

Documentation of mining damage removal and protection against the projected mining influences is performed if the object is not able to take over the predicted mining impacts or there are any significant damages of the structure, reducing its load capacity and/or resistance to mining influences. A significant part of such project is doubled with the above-mentioned resistance assessment. A designer of renovation/reconstruction must therefore obtain a lot of the same information. BIM technology would provide ready digital bridge model for such the designer. After the completion of the project, the designer should provide data for updating the BIM model on a common platform (i.e. the model available to all participants of the process). Visit on the bridge object can be used in this case to validate the BIM model. The digital bridge model can be quickly implemented into new analyzes, what significantly reduces the necessary time and cost of analytical calculations and reduces the risk of errors.

5 Incorrect correlations in investments

The mine orders making a design documentations regarding on elimination of mining damages and preventive actions and provides reconciliation of this documentations by bridge administrator/owner. Unfortunately, in many cases, bridge administrators do not inform local mines about activities in the future regarding to the object, e.g. replacement or significant reconstruction of the bridge object in the near future. In BIM, data on investment plans of managers/administrators of bridges or watercourses as well as planned activities by mines in the field of road and rail infrastructure could be described on the timeline as an additional parameter. Some examples of the lack of correlations in investments are provided below.

Three-span bridge frame has been protected against mining influences for the entire estimated mining exploitation, i.e. for about 20 years. The mine incurred significant costs. Three years after renovation, the bridge's administrator decided to demolish the object and build a new one due to insufficient load-bearing capacity of the structure and insufficient width in relation to the parameters of the modernized road [6, 7].

Mine repaired the railway viaduct in December, and in June of the following year, the corroded spans of this viaduct were replaced with new ones, and the abutments and pillars partially demolished and seriously rebuilt as part of the revitalization this PKP line [16].

Mine wanted to liquidate floodplains formed on the river as a result of the area subsidence and commissioned making documentation which allow on lowering the river bed to a specified level. It was necessary, among others, to protect the foundations against undercutting by river, as well as pillars and abutments against the loss of stability in the case of four bridges [10]. It was necessary to correct the shoreline. It took about two years to make this documentation. Documentation ultimately was not sent for implementation, because on the request of the watercourse administrator, a project to strengthen the river bed on the 500 m section at the connection of the river and the local stream was initiated, so that to prevent flooding and silt up bottom - this watercourse administrator project was part of a larger task where a large EU subsidy was obtained (the change in the "EU" project threatened to return the subsidy for the entire task) [10].

6 Technical conditions and resistance to mining influences

Static and dynamic resistance of bridge structures to mining influences is not constant, but changes over time. Mining exploitation has a significant impact, e.g. by stretching/squeezing expansion joints, introducing significant displacements into bearings [9, 12, 13, 14, 15, 16, 17]. The mining turning and tilting of abutments may lead to cracking on the abutments and spans [5, 6, 9, 11, 12, 14, 15]. The change in the curvature of the area has a negative effect on the construction of abutments - significant cracks may be created that destroy the integrity of the solid of bridgehead [8, 9, 11, 12, 15]. The uneven support subsidence, associated with the curvature, generates additional internal forces in continuous multi-span structures [6, 7]. In hyper static structures (e.g. multi-span continuous beams, frames), appears scratches which can develop into significant cracks [6, 7, 8, 9, 12].

Resistance of bridge structures to mining influences also changes as a result of normal using and impact of environmental (atmospheric) factors, e.g. bearings made of steel plates may be rusty and blocked [9, 13].

Damage controls in traditional bridge maintenance systems are often ineffective, because such controls are usually performed too rarely and are not related to the occurrence of mining influences on the surface of the area. Bridge calculation models for assessment of resistance to mining influences are rarely verified for damages. In BIM technology, damage can be included into the model. The great advantage of BIM technology is the ability to use (in order to create, validate and update the model) many modern techniques that allow remote, automatic acquisition of information about the object, damages, displacements, e.g. by using electronic sensors, drones, satellites, cameras. Especially significant possibilities, in the case of bridge objects located on a deformed area, are given by laser scanning. Laser scanning is a technology that allows on the visualization of surrounding space by using point's cloud. Model of the object is created from the point's cloud. Laser scanning can be realized from ground (from permanent or mobile positions with a device mounted on a vehicle) or as an airborne from planes or drones.

BIM enables the use of series of advanced digital technologies to assessment of technical conditions. Augmented reality (AR) is particularly interesting in the case of mining damages [e.g. 5]. During bridge inspection operator can display damages recorded on the object during the previous inspection and compare damages on tablet (in the case of larger financial expenditures, the operator can be equipped with 3D projection glasses). By comparing images it is easier to compare damages and reliably to assess the propagation of damages - reduced are the time required for a good review and the cost of human work. In case of several bridge objects in a given mining area, bridges to frequent reviews (e.g. in case of significant mining influences or objects sensitive to mining influences), this operation (i.e. AR) may be cheaper than sending several people to describe the damages, moreover the review report is performed much faster (with less work).

7 Acquiring information from monitoring systems

The basic disadvantage of traditional measurements is their low frequency, generally 2-4 measurement cycles are carried out during the year. Rarely perform measurements do not give full information about mining influences in the area of the object - stretching and compression in many situations follow each other and partly abolish, extremes of deformation and the greatest threats often appear between measurement cycles and are not documented in any way. The results of traditional geodetic are not available in digital form, so automatically implementation of measurements in computer systems is difficult. Changes in measured quantities are manually calculated and referred to the range of bearing work at a certain point of time considered to be initial, then the displacement are calculated. The next step is to determine the displacements which will be caused by another, next mining exploitation and analyzes whether the bridge bearing will transfer these influences.

The solution to the above problems is remote, frequent measurements carry out with a certain frequency (e.g. every hour). There are several pilot remote monitoring systems installed in Poland on bridge structures on mining areas [16, 17]. Bridge administration has access to measurements perform by these systems, but the lack of appropriate digital models of bridges, force manual data processing.

The starting point during projecting of monitoring system should be real object. The parameter values measure by the system should be related to the values determined earlier on the basis of the numerical model solution and on the basis of mining prediction. The measurement, without comparison of the measured value, is just a simple observation. The next stage is defining the object's resistance to mining influences, e.g. analysis of: the layout and ranges of bearings, the widths of expansion joints, the width of the bearing benches. BIM technology enables the inclusion of measurement data into the model and the use this data also automatically. After defining boundary conditions for an emergency situation, the system can automatically generate warnings, which facilitate supervision of the object and directly translate into improved safety.

8 Summary and final conclusions

The article discusses problems related to the maintenance of bridge objects on mining areas. It is pointed out that in the field of BIM technology are currently ready solutions and tools for obtaining, modeling and managing information. Advantages of BIM technology uses to optimize the process of maintaining and managing bridges are described:

- reduction of mine costs relates to easy access to a good digital bridge model, currently identical information are obtained by different authors independently (e.g. geometric inventarisation of object, identification of bearings and material data),
- increase the quality of documents on the resistance of bridge structures to mining influences and documents related to the removal of damage and bridge protection,
- actual information about the mining influences that occurred, are occurring now and are predicted in future, related to the relevant time periods,
- information about the mining influence which bridge received and can receive in future,
- the possibility of adjusting the scope of protection solutions to mining influences and the scope of repairing works related to the removal of mining damage to the plans of the bridge administrator, reduction costs of mines,
- adaptation of investment plans (e.g. repairs, maintenance, reconstruction) to current and predicted mining exploitation and associated with it costs reduction which are bearded by the administrator of the bridge,
- avoidance of accident situations, general improvement of safety,
- increased trust between the participants of the process thanks to the transparency of each actions,
- improvement of the climate for mining exploitation in region of the object, theoretically it will be easier for mine to obtain agreement for the exploitation of larger number of coal walls, get permission for larger limit values of deformation indicators,
- the ability to control the damages development and including the damages into the digital bridge model,
- improving the quality of information about the bridge objects, its technical condition and current kinematic possibilities (e.g. ranges of bearings and expansion joints), availability and easy interpretation of the results of geodetic measurements or data from automatic monitoring systems, thanks to the connection of the measurement with the information about the object and boundary conditions of the displacements.

References

1. A. Tomana, BIM Innowacyjna technologia w budownictwie. Podstawy, standardy, narzędzia (eng. BIM. Innovative technology in construction. Basics, standards, tools). Builder, PWB MEDIA, Kraków (2016).
2. D. Kasznia, J. Magiera, P. Wierzowiecki, BIM w praktyce (eng. BIM in practice). PWN, Warszawa (2018).

3. J.M. Davila Delgado, L.J. Butler, N. Gibbons, I. Brilakis, M.Z. Elshafie, C. Middleton, Management of structural monitoring data of bridges using BIM. In Proceedings of the Institution of Civil Engineers-Bridge Engineering, Vol. 170, No. 3, pp. 204-218, Thomas Telford Ltd, (2016, November).
4. B. McGuire, R. Atadero, C.M. Clevenger, M.E. Ozbek, Bridge Information Modeling for Inspection and Evaluation. ASCE Journal of Bridge Engineering Vol. 21 (4), 04015076, pp. 1-9, (2016).
5. M. Salamak, M. Januszka, Technologie BIM i poszerzonej rzeczywistości w inspekcji obiektów mostowych (eng. BIM and augmented reality in the inspection of bridge structures). Mosty, nr 5, pp. 28-35 (2016).
6. P. Bętkowski, S. Pradelok, M. Łupieżowiec, Maintenance and risk assessment of a concrete frame bridge impacted by mining deformations of the area. 14th Inter. Multi. Scien. GeoConfer. SGEM 2014, Conf. Proceedings, Vol. 3, pp. 345-352 (2014).
7. P. Bętkowski, S. Pradelok, M. Salamak, Concrete frame bridge impacted by mining deformations. The 9th Cent. Eur. Cong. on Concrete Engineer., pp. 192-195 (2013).
8. P. Bętkowski, Conclusions from observation the small crisp arch bridges located on mining areas. WMCAUS 2016. Procedia Engineering, vol. 161, pp. 687-692 (https://doi.org/10.1016/j.proeng.2016.08.738) (2016).
9. P. Bętkowski, Repair and protection of small railway viaduct with jammed span at the mining influence. IOP Conference Series: Materials Science and Engineering, vol. 245 022018 1757-8981, pp. 1-10 (doi: 0.1088/1757-899X/245/2/022038) (2017).
10. P. Bętkowski, Dostosowanie obiektów mostowych do obniżonego koryta rzek i potoków (eng Adapting of bridge objects to the lowered channel of rivers and streams). Przegląd Górniczy, nr 8 (70), pp. 65-70 (2014).
11. P. Bętkowski, Przebudowa i zabezpieczenie na wpływy górnicze niewielkiego wiaduktu kolejowego zintegrowanego z nasypem (eng. Rebuilding and protection to mining influences the small railway viaduct integrated with the embankment). Przegląd Górniczy, nr 3, pp. 1-7 (2015).
12. P. Bętkowski, Przebudowa wiaduktu kolejowego będącego pod wpływem znacznego górniczego osiadania terenu (eng Rebuilding of the railway viaduct being under the influence of considerable mining subsidence of area). Przegląd Górniczy, nr 12 (72), pp. 100-106 (2016).
13. P. Bętkowski, Wybrane problemy łożyskowania starych mostów kolejowych położonych na terenach górniczych (eng. Selected bearing problems of old railway bridges located on mining areas). Wiadomości Górnicze, nr 5, pp. 262-270 (2017).
14. A. Rosikoń, Budownictwo komunikacyjne na terenach objętych szkodami górniczymi. WKŁ, Warszawa (1979).
15. M. Salamak, Obiekty mostowe na terenach z deformującym się podłożem w świetle kinematyki brył (eng. Bridge structures located on areas with ground deformations in the light of solids konematics). Wydawnictwo Politechniki Śląskiej, Gliwice (2013).
16. P. Bętkowski, Ł. Bednarski, R. Sieńko, Structural health monitoring of a rail bridge structure impacted by mining operation. Technical Transactions Civil Engineering, 6-B (i.21), Wydawnictwo Politechniki Krakowskiej, Kraków, pp.15-27 (2014).
17. B. Parkasiewicz, M. Kadela, P. Bętkowski, R. Sieńko, Ł. Bednarski, Application of structure monitoring systems to the assessment of the behaviour of bridges in mining areas. IOP Conference Series: Materials Science and Engineering, vol. 245 032018 1757-8981, pp. 1-10 (doi: 10.1088/1757-899X/245/3/032018) (2017).

Effects of strong mining tremors, and assessment of the buildings' resistance to the dynamic impacts

Izabela Bryt-Nitarska[1,*]

[1]The Strata Mechanics Research Institute PAN, Cracow, Poland

Abstract. A particularly important element in the assessment of the actual state of the threats which is caused by conducting the mining exploitation of seams bumping under the urban areas is to diagnose the condition of the land development after hard shocks. In the buildings, for which the impact of the mining activities, including the tremors, is not taken into account at the stage of design and formulation of the strength and use conditions, conclusions from the structure behaviour under the tremor influence are an essential part of the assessment of the possibility for transferring the further dynamic impacts. The use of conclusions from the *in situ* research has its role in anticipating the behaviour of the buildings in case of the forecast of the mining tremors effects in the regions of their impacts. These conclusions should also provide ground for the assumptions to the scope of the building prevention necessary to be taken in case of forecasting the tremors with big intensity. Based on the analysis of effects which occurred in the land development after the high-energy mining tremors, the elements of the dynamic resistance assessment for the buildings with traditional structure were discussed.

Keywords: dynamic resistance of the building, effects of mining shocks, strong mining shocks

1 Introduction – concept of the resistance to mining impacts

The mining impacts often cause deformations and damage in the building structure, that are unacceptable in the assessment of the standard ultimate limit states (ULS) and serviceability limit states (SLS) [1, 2]. The possibility to accept, from time to time, the reduced requirements in the degree of meeting the SLS in the cubature objects subject to the mining impacts is practiced though. This approach is underpinned by the conclusions from the studies over the real behaviour of the structures under the conditions of load impacts derived from the mining area deformations [3]. While assessing the effects formed in the structure, a prerequisite is, however, the knowledge on the values of measuring indicators describing the occurred area deformation. With regard to the object it is required

* Corresponding author: nitarska@img-pan.krakow.pl

to know the design and material conditions as well as the technical condition. The assessment of the construction object is conducted in the context of expected category of the mining area and foreseen dynamic impacts.

According to the definition formed in this paper [4], an object resistance is understood as capability for shifting the mining impact, expressed in the units of that impact, while not exceeding a specific technical condition of that object at the same time. With reference to the dynamic impacts, it means the possibility to accept additional forces formed in a building as a result of mining tremors [5]. The concept of such understood 'object resistance' is made for the procedure purposes enabling to carry out the exploitation of mineral deposits within the operation of underground mines under urban areas.

2 Conditions for dynamic resistance of buildings

For the brick buildings the resistance to the dynamic impacts, triggered by mining tremors in the ground base, determine mainly tensile and shear strength of the wall. The wall strength is a function of material features attributed to masonry units and binder. In the objects that are long-term used, the wall strength is affected also by the phenomena of shrinkage, creep and beyond-elastic strains in the strenuous parts of the structure [6, 7]. A key determinant for the level of the real wall strength is the aging process related to the technical wear and the fatigue process resulting from multiple exposures to the dynamic loads [8, 9].

An impact resistance is also affected by the building mass spatial stiffness defined by non-dilatation strain [10], specific symmetry of the structure shape and the mass distribution uniformity from the self-weight loads. Elements responsible for the spatial structure stiffness are the applied system of stiffening walls, upholding the horizontal projection symmetry, stiffness of ceilings and their connection with the structure. In the buildings with wooden ceilings, for which the horizontal stiffness decrease is characteristic, an element determining the wall strength is also the wall stability (buckling) [5, 10, 11].

As a result of the dynamic loads in the structure there are generated [6]:

- bending from the plane, for the walls located perpendicularly to the wave propagation direction,
- bending in the plane, for the walls located parallel to the wave propagation direction,
- shearing from the wall plane and in the wall plane.

In the existing buildings, the criterion of 'shock resistance' is the wall shear resistance and the horizontal member resistance. For the uniform wall, it is the load capacity in the corners and on the wall contacts as well as the load capacity in the places where the stiffness is impaired by openings. In the horizontal contacts, in the places where the slab roofs, ceilings and tie beams are supported on the wall, and the opening lintels, ceiling bearing beams and staircase beams. The criterion of 'shear resistance' refers also to free sections of masonry structure such as gable walls, attics, chimneys.

The wall shear resistance induced by horizontal inertia forces depends (directly proportional) on the degree of compressive stresses occurring in the strenuous cross-sections [6, 7]. In case of the wall subject to the shear parallel to the support joints, the exceeding of the wall resistance consists in the occurrence of 'a slip' in the support joint or on the contact of joint and bricks. Such damage arises in the wall structure with the mortar shear strength lower than the masonry units' tensile strength. In case when the mortar in the masonry structure is characterised with the shear strength bigger than the masonry units' tensile strength, and at the bigger compressive stresses, perpendicular to the support joints' surface, an oblique scratch appears in the wall.

 https://doi.org/10.1051/e3sconf/20183601003

In practice, the structural system resistance depends on the current conditions within the range of real redistribution of internal forces related to the masonry structure condition, in particular, occurrence of wall geometrical discontinuity in form of scratches and cracks.

3 Observed effects of the mining tremor impacts

The presented examples of damage occurred in the land development which was subject to the impacts of many high-energy mining tremors during the last sixteen years. The tremors were generated by the exploitation realised in the seam characterised with high seismic activity. Over 100 tremors of energy around 10^6 J were recorded in the exploitation area. The land development is also located within the range of impact of four incidental tremors with energy $7x10^7$ J ÷ $8x10^8$ J, which occurred over a short period of time. The distance of the buildings from the tremor epicentre was each time from ca. 7 m to 1,000 m. The measured maximum surface vibrations velocity might have been up to ca. 0,037 m/s. From the obtained distribution of the maximum amplitudes of the ground vibration acceleration resulted that the buildings might have been subject to the impact of the accelerations exceeding 1,2 m/s^2 and the value of maximum acceleration might have been ca. 1,9 m/s^2 [8, 9].

The diagnostic measures were based on the inspections of the building structure damage taken after the strong tremors. In some buildings the author made the inspection three times over the 16 years. It enables to conduct the comparative analysis of the damage after the tremor with the previously recorded damage.

The buildings suffered a great deal of damage in the structural members. In the brick walls of a two-storey building stiffened with horizontal in-plane loaded plate over the ground floor, the scratches appeared, that are characteristic for torsional-flexural work of the wall corner and shear of wall derived from a variable as for the horizontal force direction – fig. 1 and 2 (acc. to [8]).

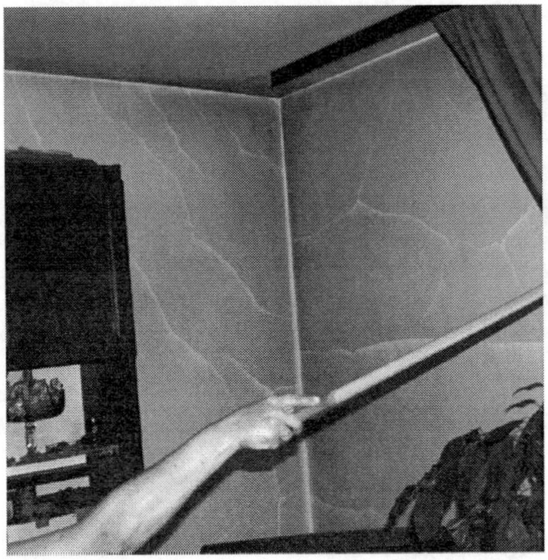

Fig. 1. Wall scratches in the area of the outer corner of the building

Fig. 2. Wall scratches in the area of the outer corner of the building

In the buildings with wooden ceilings which do not provide enough stiffening for load-bearing walls, the damage accumulated in the walls between windows. In many cases strong damage of lintels was observed, which under the conditions of multiple impact of tremors lead to disintegration of the masonry structures and require to be reconstructed – fig. 3 and 4.

Fig. 3. Visible disintegration of the wall by the window

Fig. 4. Cracking and displacement of the lintel of the window

In the line of buildings erected in the industrial construction technology there were numerous horizontal shears of internal cross and long walls. Despite the fact that the walls are made of concrete, anti-shrinkage reinforced, in the places where the stiffness was impaired with the communication passage openings, strong cracks were made – fig. 5 and 6.

Fig. 5. Horizontal cracking of the concrete inner wall

Fig. 6. Horizontal cracking of the concrete dilatation wall

For multi-segment development the criterion of resistance to the dynamic impacts includes also the width and passability of the maintained expansion gap. Improperly executed or filled expansion gaps cause the structure damage intensification in areas close to the expansion gaps.

A special case represents the damage of the concrete ceiling made with an economic method, while making a superstructure over a single storey building. The morphology of damage was managed to be analysed carefully only after hacking off the plasters. The horizontal cuts formed along the edges of the ceiling support and the scratches on the bottom plate surface result from the lack of properly shaped tie beam whose role is to provide restraint of the ceiling. Due to the dynamic impacts, the workmanship irregularities caused the formation of cracks in the support and span cross-sections – fig. 7 and 8.

Fig. 7. Cracking of a reinforced concrete floor slab.

Fig. 8. Cracking of a reinforced concrete floor slab based on.

The damage was commonly stated in the finishing elements i.e. in the ceramic wall and/or floor linings, in the contacts of non-bearing walls e.g. gypsum plasterboards. In the buildings subject to several inspections, the intensification of the previously observed damage was found. The intensification consisted in increasing the openings of scratches or/and their length in the structural members and in the finishing elements.

4 Determinants of the dynamic resistance assessment

In the real conditions of impacts, the mining tremors cause the formation of torsional-flexural loads in the structure. The building parts are subject to variable displacements on a cyclical basis in plane and from plane forced by the structure torsion. The varied strength properties of the wall cause that the wall structures of the small-size elements are very susceptible to damage induced by the mining-origin tremors. In the potential risk assessment for occurrence of the damage that may threat local capacity or stability of the structure, the fact should be taken into account that due to strong tremors and multiple impacts of the dynamic load the arising horizontal forces may lead to the disintegration of the masonry structures. The multiple impacts of the dynamic loads substantially decrease the structure resistance to the vertical loads as well as to subsequent loads with horizontal forces. When the scratches appear in the ceramic linings, facade plasterwork or on the contacts of gypsum plasterboards under the conditions of multiple impacts of the dynamic loads, it indicates possible temporary overloads of the masonry structures. Whereas the scratches appearing on the contacts of prefabricated members, in the situation of tremor impact, may indicate the occurrence of excessive forced displacements of the structure.

In case when a building is located in the area of elevated seismic activity, an impact of single tremor on the structure condition cannot be considered. Since the whole history of loads decide on the condition of damage. Every subsequent tremor impact on the structure, particularly on those less resistant, causes its strength deterioration. Then, structure damage may appear even in the case when the impact of tremors is of the smaller intensity than the previous ones.

5 Summary

The assessment of structure strength to the mining loads should be always considered while taking into account the necessity for meeting the demands as for the basic requirements for the buildings, pursuant to Art. 5. p. 1 of the Act [12]. In mining regions, they are as follows: structure safety, serviceability safety and protection against (...) vibrations, mining tremors.

In no event of actual implementation of shifting additional mining loads, even while considering temporary reduced serviceability requirements, should be allowed to exceed the structure safety limit. In the assessment methodology for the building objects' dynamic resistance we cannot ignore the fact that dynamic phenomena of high intensity occur several times during their use. As a result of those impacts the structures suffer often minor damage which causes redistribution of forces and stresses as well as changes in the dynamic structure reaction conditions. The dynamic resistance of the buildings, over the course of time and occurrence of further dynamic phenomena, is declining.

In the near future we may encounter the real necessity for taking, more frequently than so-far, the reinforcements of the existing structures to the dynamic impacts.

References

1. PN-EN 1996-1-1 Eurokod 6. Projektowanie konstrukcji murowych.
2. PN-ISO 2394:2000 Ogólne zasady niezawodności konstrukcji budowlanych.
3. F. Pachla, T. Tatara: Odporność zabudowy mieszkalnej i gospodarczej na wpływy eksploatacji górniczej w obszarze górniczym KWK Janina. PAN (2017), nr 101, (str. 45-60).
4. J. Kwiatek, Obiekty budowlane na terenach górniczych. GIG, (2007).
5. B. Gil-Kleczeńska, J. Kwiatek, L. Muszyński. OTG, 71/1 (1985).
6. B. Lewicki, R. Jarmontowicz, J. Kubica: Podstawy projektowania niezbrojonych konstrukcji murowych. ITB, (2001).
7. B. Lewicki, Mat. Bud. no. 4/1999.
8. I. Bryt-Nitarska, GIG, no. 2/1/2011 (p. 75÷84).
9. I. Bryt-Nitarska. Miernictwo górnicze i ochrona terenów górniczych w obecnych warunkach wydobycia złóż surowców mineralnych w Polsce. AGH, 25-34 (2016).
10. J. Fedorowicz, M. Kawulok, GIG, no. 20 (1997).
11. E. Maciąg:Problemy Projektowania i Ochrony Obiektów Budowlanych na Terenach Górniczych, ITB, 1999.
12. Ustawa z dnia 7 lipca 1994 r. Prawo budowlane (Dz. U. poz. 1332).

The requirements for implementing Sustainable Development Goals (SDGs) and for planning and implementing Integrated Territorial Investments (ITI) in mining areas

Lucyna Florkowska[1,*], and *Izabela* Bryt-Nitarska[1]

[1] Polish Academy of Sciences, The Strata Mechanics Research Institute, Cracow, Poland

Abstract. The notion of Integrated Territorial Investments (ITI) appears more and more frequently in contemporary regional development strategies. Formulating the main assumptions of ITI is a response to a growing need for a co-ordinated, multi-dimensional regional development suitable for the characteristics of a given area. Activities are mainly aimed at improving people's quality of life with their significant participation. These activities include implementing the Sustainable development Goals (SDGs).

Territorial investments include, among others, projects in areas where land and building use is governed not only by general regulations *(Spatial Planning and Land Development Act)* but also by separate legal acts. This issue also concerns areas with active mines and post-mining areas undergoing revitalization. For the areas specified above land development and in particular making building investments is subject to the requirements set forth in the *Geological and Mining Law* and in the general regulations. In practice this means that factors connected with the present and future mining impacts must be taken into consideration in planning the investment process.

This article discusses the role of proper assessment of local geological conditions as well as the current and future mining situation in the context of proper planning and performance of the Integrated Territorial Investment programme and also in the context of implementing the SDGs. It also describes the technical and legislative factors which need to be taken into consideration in areas where mining is planned or where it took place in the past.

Keywords: mining areas, mining damage, buildings, Sustainable development Goals (SDGs), Integrated Territorial Investments (ITI)

* Corresponding author: florkowska@img-pan.krakow.pl

1 Introduction

The effectiveness of a socio-economic policy in achieving economic growth goals greatly depends on being able to properly assess and make provisions for local conditions, which can have a significant influence on shaping the environmental, technical, economic and social aspects of life in a given area. This is why local conditions should be used to determine the framework and directions for proper task planning and implementation.

Mining areas – rich in natural resources, and for this reason subject to strong anthropopressure – are regions of very specific nature. This special character can be seen in many areas of socio-economic life. This paper points out and describes the specificity in terms of geotechnical and structural conditions of planning, erecting and using buildings in the aspect of compatibility with the sustainable development goals and of SDGs implementation tools.

The specific nature of the geotechnical and technological requirements for erecting buildings in mining areas has a significant impact on the technical and economic aspects of construction projects. This special nature also applies to the safe use of buildings and infrastructure. For this reason appropriate shaping of a socioeconomic policy requires the use of research findings in elaborating and implementing procedures which impose both the obligation to use appropriate geotechnical and structural solutions in building construction and the obligation to exploit deposits in such a way that damage is minimized and failures and disasters can be prevented.

1.1 Sustainable Development Goals (SDGs)

Managing natural resources in a way that minimises the damage caused by their use, limits the transformation of natural ecosystems and saves some of these resources for future generations is not only an ethical but also a legal requirement. Sustainable development is generally accepted as the idea of achieving a harmonious improvement in the standard of living and scientific, technical and economic advancement at the same time protecting the environment and natural resources and bearing in mind the needs of future generations. The idea of sustainable growth has been transformed from a theoretical doctrine into a system of legal regulations which aim to shape social and economic changes in the world [1].

In September 2015 the international Sustainable Development Summit was held in New York, whose goal was to develop a new, global development strategy until 2030. This strategy was supposed to continue, but also to expand, the activities undertaken to achieve the Millennium Development Goals (MDGs) adopted in UN Millennium Declaration in 2000. One hundred UN member states, including Poland, signed the summit's final document called "Transforming Our World: the 2030 Agenda for Sustainable Development".

The 2030 Agenda lists 17 main Sustainable Development Goals (SDGs) with allocated operational targets specifying ways of achieving them (fig. 1). They make up a list

Fig. 1. Sustainable Development Goals - SDGs (http://www.un.org/sustainabledevelopment/sustainable-development-goals/)

of 169 agreed global solutions, which should be reflected on national and regional levels in actions undertaken with local conditions in mind.

Poland, as a signatory of the Agenda, is obliged to make efforts to achieve the set goals, adjusting the scope of actions undertaken to the situation in respective regions. In Poland legal requirements concerning the implementation and enforcement of sustainable development principles are regarded as important. Article 5 of the Constitution of the Republic of Poland states that the Republic of Poland shall follow the principle of sustainable development. The national sustainable economic, environmental, social, administrative and political development will be implemented through relevant statutory regulations. Also, a number of strategic documents were drawn up, which determine the so called development policy as well as the principles of monitoring its progress [2].

1.2 Integrated Territorial Investments (ITI)

Integrated Territorial Investments are one of the tools used to implement sustainable development goals. They are mainly concerned with sustainable development of urban areas. Both because of the areas where they are to be implemented and the kind of actions to be undertaken, ITI can be especially beneficial in mining areas. The area of ITI activities overlaps with areas affected by mining, which are usually also highly industrialized and urbanized. In Poland the Upper Silesian Coal Basin is one such area, where large urban agglomerations are affected by the impact of repeated underground mining operations and where numerous degraded areas can be found. According to a document called *"The principles of implementing Integrated Territorial Investments in Poland"* (The Ministry of Regional Development, Warsaw 2013) one of the main directions of actions undertaken under the ITI programme is *"Restoring the socio-economic functions in devastated urban areas"*. The first ITI-funded projects were carried out in the Silesian region. The ITI initiative is being implemented there in the so called Central Sub-region, where funding was granted among others to brownfield area revitalisation projects. Restoring functional value to devastated areas, especially in the context of turning post-mining areas into investment areas, is a problem which requires specialist knowledge. To carry out such undertakings effectively, solutions should be in place to ensure the permanence of the results, the safety of buildings erected in the revitalized area and measures against increased wear of buildings and infrastructure. This is why the procedures of granting funding under the ITI framework should include the requirements of carrying out relevant tests of the state of the subsoil, analyses of the geotechnical and mining situation and of applying appropriate technical solutions. To ensure that the above requirements are fulfilled project funding applications should be verified by experts who have the knowledge of the impact that subsoil deformations can have on buildings. Due to local conditions in which additional loads are generated by mining any construction project in mining areas must fulfil additional requirements. Meeting these requirements should be obligatory for any investment project carried out as part of the ITI programme.

2 The specific character of mining areas

The specific character of mining areas can be seen in many aspects of social and economic life. The characteristic feature of mining areas is the so called mining damage and the existence of degraded areas of no functional value [3,4].

The term *mining damage* is used to describe damage caused by the operation of a mining facility. In practice, this term is used mainly in the context of changes to the morphology and hydrological conditions, of dynamic events with the characteristics of tremors and of the impact these processes and phenomena have on buildings [5,6].

However, in the light of the above definition, negative changes in the environment (including changes in air quality) caused by dust and gas emission to the atmosphere as well as water and soil pollution (including salinity) which accompany the exploitation of solid minerals and fluidized deposits should also be treated as mining damage. These direct effects of mining bring about wider economic, social and environmental phenomena known as the effects of mining damage (Fig. 2)[4].

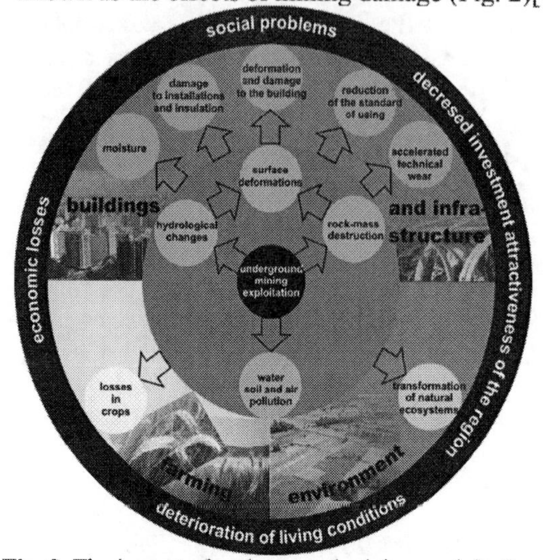

Fig. 2. The impact of underground mining exploitation on the surroundings [4]

In the material aspect negative effects of mining manifest themselves as:

- the destruction of rock mass structure in the proximity of the excavation,
- surface deformations above the excavation,
- hydrological changes,
- deformations and damage to buildings, civil engineering structures and technical infrastructure,
- an increased wear of buildings, civil engineering structures and technical infrastructure,
- air and soil pollution.

Examples of significant damages created in building construction elements due to the effect of loosening of a ground caused by mining operations are presented in Fig. 3. In the cases presented, there was a loss of load-bearing capacity of the construction elements, combined with a threat to the safety of building. These damages were revealed in the buildings due to the movement of the mining operations front beneath them.

Fig. 3. Cracks in load-bearing walls due to loosening of a ground caused by mining mining floor dislocations [4]

https://doi.org/10.1051/e3sconf/20183601004

Fig. 4a shows an example of a building that was in the range of the impact of the outcrops of the active fault of the geological layers. Activation of the fault zone was triggered by mining exploitations, which in several subsequent seams undermined the dislocation plane. As a result of repeated exploitation of the deposit in the vicinity of an active geological fault, in the building the damage occurring during subsequent years led to the necessity of excluding it from use and for demolition, despite taking protections and repairs (Fig. 4b)

Fig. 4. Damage of a building on the convex periphery of the mining trough, while the impact of terrain threshold a) cracking of the front wall - condition for 2002, c) condition of the building in 2014

From the kinds of mining damage mentioned, the damage which occurs in buildings and infrastructure is definitely the most apparent and causes the greatest problems for the people living in areas affected by mining. Deformations and damage to buildings, roads and utilities (power, gas, water supply, sewerage, heating and communications) cause problems with using them and/or put the safety of these structures at risk bringing about economic and social consequences both locally (owners and users) and globally - influencing the general condition of the area and its inhabitants.

Material mining damage causes unavoidable economic consequences. These consequences affect both the mining company itself but also the owners and users of property and buildings exposed to mining impact. The cost of protection and strengthening devices along with minor and major repairs are the main economic consequences of mining damage for property owners and users. A general economic consequence is the property's decreased value due to both increased technical wear of the building and the fact that the property stands on land threatened with further ground deformations [4,7].

4 Local technical requirements for mining areas

From the point of view of planning and carrying out technical projects, the key fact is that in areas exposed to the impact of mining the structural components of buildings must bear additional loads caused by deposit exploitation. The additional requirements resulting from this have been imposed by relevant legal regulations. As early as at the project planning stage a special approach to design making is necessary for mining areas. According to the Decision of the Minister of Transport, Construction and Maritime Economy *on the detailed scope and form of building permit design* of 25 April 2012 (Journal of Laws of 27 April 2012, item 462) the site development plan (which is an integral part of the building permit design) in its descriptive part should contain *"data specifying the impact of mining on the site of the planned construction undertaking situated within the borders of a mining area"* (§ 8.2, p.6), and the technical description of the design should include *"the geotechnical category of the building/structure, the conditions and method of laying its foundations and of securing it against the impacts of mining"* (§ 11.2 p.4).

One of the fundamental aspects of design making is determining the geotechnical category of the building. According to the Decision of the Minister of Transport, Construction and Maritime Economy *on the detailed scope and form of a building permit design* of 25 April 2012 (Journal of Laws of 27 April 2012, item 463) on determining the geotechnical conditions of laying foundations of buildings/structures complicated subsoil conditions exist *"in the case of subsoil strata affected by adverse geological phenomena, in particular processes and forms connected with carstification, landslides, scouring, quicksand, glacial tectonics, expansive and subsidence-prone soils; situated in mining damage areas with possible discontinuous rock mass deformations, in river valleys and deltas as well as in sea areas"*.

According to the classification used in Eurocode 7 (PN-EN 1997 Geotechnical designs) a building erected in mining areas belongs to the third geotechnical category, which comprises among others *"buildings/structures in areas with a high probability (...) of long-term subsoil movements"* (chapter 2.1).

According to the Decision of the Minister of Infrastructure of 12 April 2002 *on the technical conditions that buildings and their location should meet* (Journal of Laws no. 75, item 690) mining-induced damage belongs to negative impacts which restrict the possibility of erecting buildings intended for human occupancy in areas affected by it; § 11.1 sets forth that the condition which must be fulfilled if a building is to be erected in the area affected by such negative impacts is the use of technical means mitigating the negative effects or increasing the building's resistance to these impacts. *"Measures to secure the structure of a building appropriate for the type of threat resulting from anticipated mining impact, which is understood as constrained displacements, deformations and tremors of the subsoil must be used in areas affected by mining impact"* (§ 205). The Decision also contains an exception concerning the fulfilment of the serviceability limit states (SLS) criteria for *"these deformations, damage and tremors of the structure which result from impacts caused by mining"* (§ 204.6).

To properly design protections for a building at risk of mining damage additional construction requirements must be fulfilled to meet the requirement of securing the structure against dangerous subsoil deformations. When designing a building situated in a mining area, one must also remember to meet all the requirements specified in relevant regulations. When formulating the assumptions for calculations it is worth bearing in mind that:

1. The values of land deformation indexes which are found in a forecast refer to the state which is the state of the surface at the moment immediately preceding making the forecast and are valid only for the part of mining included in the mining operation plan (MOP) for several years.
2. In the vast majority of mining areas mineral exploitation has lasted for a long time, which has caused the original rock mass structure to be significantly disturbed, and can increase the extent and magnitude of surface deformations.
3. When designing a building in an area where many panels were mined in several (sometimes more than ten) seams, it is a mistake to omit this fact and fail to recognize the possibility that old goaf can be activated or mining effects can overlap [8].

5 Summary

When situating economic undertakings in mining and post-mining areas important local conditions must be taken into consideration. They play a special role in construction projects and in revitalization projects to restore degraded areas. As areas where deposit mining takes place are especially under threat due to strong anthropopressure, it is very important to properly recognize and make allowances for local conditions and requirements for sustainable development actions to be effective.

Meeting additional requirements resulting from surface and development protection requirements is in the interest of the mining municipalities, property owners, investors and mining enterprises. A regular implementation of structural and geotechnical protection to secure buildings erected in mining areas will minimize the risk of failure and disasters, reduce an increased technical wear of buildings and minimize disputes and conflicts. The purpose of additional requirements in terms of using an extended assessment of the geological and mining situation, adjusting construction solutions to the stricter requirements, providing appropriate structural and geotechnical protection or installing building monitoring systems is to reduce the extent of the damage and at the same ensure better durability of undertakings (Fig. 5). This is why additional steps should be taken for these requirements to become obligatory in mining areas, for example by incorporating them into relevant areas at the level of making a zoning plan and in the case of state or EU-funded projects, including those carried out under the ITI framework.

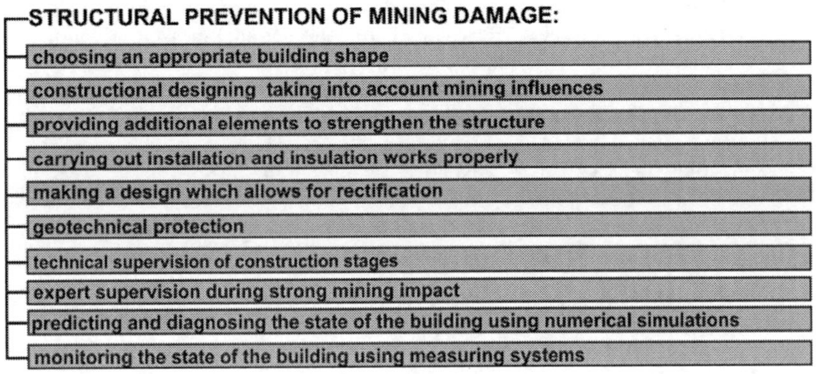

Fig. 5. Elements of structural prevention as important additional local requirements for mining areas [4]

References

1. K. Górka, Gospodarka w Praktyce i Teorii. I. 3(36), pp. 34-50 (2014)
2. L. Florkowska, Zesz. Nauk. I GSMiE PAN, pp. 105-117 I. 94, (2016)
3. L. Florkowska, Arch. of Min. Sc., V. 55, pp. 691-701 (2010)
4. L. Florkowska, I. Bryt-Nitarska, A. Maj, Mining damage in buildings. Kraków: Arch. of Min. Sc. Monography. I. 17. (2016)
5. L. Florkowska, I. Bryt-Nitarska, Przegląd Górniczy, 71(11), pp. 101-105 (2015)
6. O. Kaszowska, "The Environment protection in mining areas". pp. 189-209 (2006)
7. I. Bryt-Nitarska, Zesz. Nauk. I GSMiE PAN, issue 94, pp. 119-125 (2016)
8. L. Florkowska, J. Walaszczyk, J. Cygan, Prace IMG PAN, T.14, I.1-2, p.93-102(2012)
9. Transforming our world: the 2030 agenda for sustainable development. 2015. A/res/70/1 United https://corporate-citizenship.com

E3S Web of Conferences **36**, 01005 (2018) https://doi.org/10.1051/e3sconf/20183601005
BIG 2018

Industrial chimney monitoring - contemporary methods

Olga Kaszowska [1], *Piotr* Gruchlik [1] and *Wiesław* Mika [1]

[1]Główny Instytut Górnictwa (Central Mining Institute), Katowice, Poland

Abstract. The paper presents knowledge acquired during the monitoring of a flue-gas stack, performed as part of technical and scientific surveillance of mining activity and its impact on industrial objects. The chimney is located in an area impacted by mining activity since the 1970s, from a coal mine which is no longer in existence. In the period of 2013-16, this area was subject to mining carried out by a mining entrepreneur who currently holds a license to excavate hard coal. Periodic measurements of the deflection of the 113-meter chimney are performed using conventional geodetic methods. The GIG used 3 methods to observe the stack: land-based 3D laser scanning, continuous deflection monitoring with a laser sensor, and drone-based visual inspections. The drone offered the possibility to closely inspect the upper sections of the flue-gas stack, which are difficult to see from the ground level.

Keywords: building object, safety, mining area, observations, measurements

1 Introduction

Over recent decades, intensive development has been observed in the area of measuring instruments and appliances and the related advancement of measuring methods. They are used to observe surface deformation and the deformation of buildings and structures caused by underground mining. The methods include: terrestrial (land-based) and airborne laser scanning (Light Detection and Ranging – LiDAR), GPS measurements, and interferometric synthetic aperture radar measurement (InSAR). Sensors, inclinometers, tensometers and lasers are used to observe deformations of land surface and buildings within an area impacted by mining. It should be emphasized, however, that the new technologies are encumbered with certain limitations, as are traditional ones. By using them, much more data can undoubtedly be obtained, and this technology can be used for continuous measurements or monitoring. However, not all of these technologies yield sufficiently accurate data.

Despite the impressive development of measuring technologies, classical methods are still in use. Experience shows that geodetic methods continue to be indispensable whenever highly accurate coordinates need to be obtained for land surface points, buildings, structures, and changes thereof.

The decision on which technologies are to be used in specific conditions should not be driven by the method being “new” or “old”, but should be based on the potential to accomplish the goal of the measurement, i.e. to obtain data of the required accuracy. The optimisation of the observation also includes its costs.

This paper presents an interesting problem, a solution to which was sought in the course of surveillance performed by the Central Mining Institute. Due to the specific features of the chimney, the optimal decision was to use several measuring technologies and perform remote observations of its technical condition.

Surveillance of the 113-meter chimney includes checking its deflection and technical condition. Periodic measurements of the chimney deflection were accomplished using traditional geodetic methods. The Central Mining Institute used 2 methods for checking the chimney geometry (primarily its deflection): land-based 3D laser scanning, and deflection monitoring using a laser sensor. Additionally drone-based visual inspections were made. Using a drone enables the ongoing assessment of the technical condition of the chimney, with no need to make use of climbers.

This paper describes the knowledge gained from comprehensive observations of the high flue-gas stack using classical geodetic methods and new technologies.

2 Chimney description

The chimney was erected in the first half of the 1940s and it serves to discharge hot (122 - 170°C) flue gas from coke-fired and hard-coal-fired boilers. Its current height is 113 meters. Originally it was 115 meters high but in 1997 it was shortened by 2.05 meters. It is a ceramic chimney with a circular cross-section (Fig. 1, 2). The chimney diameter is 7.64 m at ground level, and 3.5 m at flue-gas outlet level. The shaft is made of stack brick, and the inner lining, of up to 40 meters, of clay bricks. The chimney's foundation is a reinforced-concrete circular plate with a diameter of 17.0 m, positioned 5 m below ground level. It is adjacent to a boiler house (Fig. 10) but they are separated by an expansion joint.

Before the mining activity started, the chimney had displayed considerable vertical deflection. Due to the potential increase of the deflection as a result of the planned mining operations, the chimney was placed under surveillance.

Fig. 1. View of the chimney

Fig. 2. The chimney location in respect of the mining plots and iso-lines of surface subsidence measured in the period of 2012 – 2016

3 Geodetic measurements

The boiler house chimney was subjected to the impacts of hard coal mining activity conducted during the period of 2013-2016 in three coal seams. Due to the specific conditions of building and structure protection, an experimental system of drift mining was used. Fig. 2 shows the location of the chimney and the mining operations.

3.1 Measurement of surface deformations caused by mining activity

Geodetic measurements of surface deformation started in September 2012. They included the levelling of dispersed points and levelling and length measurements along the lines shown in Fig. 2. The last levelling was made at the end of 2016.

The stabilised points of line 1 and line 4 are closest to the stack. Subsidence of approximately 0.16 m occurred within the chimney area. Interpolation of the measurement results led to the delineation of surface deflection iso-lines (Fig. 2) and calculation of land inclination increase. At the chimney site, the increase is 0.5 mm/m SE. At the height of 113 m, i.e. the flue-gas outlet, the deflection changes by 5.6 cm. However, one must remember that the deflection of buildings and structures does not always correspond to land inclination changes.

Additionally, calculations were made using the Budryk-Knothe method, applied by Polish coal mines to predict the impact of mining activity. The parameters in this model were determined by matching theoretical profiles of subsidence troughs and those measured along the observation lines.

The theoretically calculated subsidence within the area of the chimney is 150 mm and the inclination is 1.4 mm/m. If the chimney deflection corresponded exactly to the surface inclination, the theoretical increase of the deflection at flue gas outlet level would be 15.8 cm.

3.2 Measurement of chimney deflection

The initial geodetic measurement was performed before the commencement of mining, in April 2011. All observations showed deflation of the structure at 55 – 64 cm in a north-eastern direction (Table 1). The azimuth was 46 - 67° (computed clockwise from the north). The resultant deflection according to the last measurement of September 2017 is 56 cm, azimuth 63°. The mean error of deflection determined using this method is ±3.0 cm.

The measurements were made by determining the coordinates of the chimney centre at two levels: the ground level and the flue gas outlet level. Determination of the location of the upper section poses no problems, but the lower section is accessible only at half of its perimeter, as it is adjacent to the building. During the first measurement, coordinates of the centre at ground level were determined from the inside of the chimney during a break in its operation. However, breaks in chimney operation are too infrequent to make simultaneous measurements of its base and the upper section viable.

Chimney deflection was computed using the coordinates of the bottom part's centre determined in 2011 and the current coordinates of the upper section's centre. This means that the deflections determined using geodetic measurements are subject to a non-random error. Undoubtedly, the mining activity causes not only displacement of the chimney at the flue gas outlet level, which results from inclination and dislocation of the area surface, but also dislocation of its base. For the purpose of calculation, the coordinates determined in the first measuring cycle were taken into account, so the dislocation of the base centre was disregarded. Since determination of this figure was not technically feasible, other methods were used to measure the chimney's deflection.

Table 1. List of chimney top deflection values measured by geodetic methods

Item	Measurement Date	Deflection of chimney top [cm]	Direction of deflection []
1	Apr 2011	57	49°51'
2	Oct 2011	56	52°24'
3	Dec 2011	55	55°49'
4	Apr 2012	57	55°21'
5	Oct 2012	59	55°34'
6	Apr 2013	55	52°16'
7	Oct 2013	60	56°56'
8	Apr 2014	55	49°32'
9	Oct 2014	56	51°43'
10	Mar 2015	56	50°47'
11	Jul 2015	57	50°47'
12	Sept 2015	58	46°01'
13	Apr 2016	58	51°31'
14	Sept 2016	64	55°27'
15	Oct 2016	63	57°00'
16	Apr 2017	57	67°03'
17	Sept 2017	56	63°21'

4 Continuous measurement of deflection

In the period from March to December 2016, the chimney underwent continuous measurements of deflections, using a laser sensor constructed at GIG [1]. The sensor and the recording device were installed on the chimney base wall approximately 2m above the ground level (Fig. 3a).

The sensor is based on a module of a vibration and deflection sensor with the following measuring parameters:

– range of measurement: 5 mm/m (possible adjustment to 2 to 50 mm/m),
– resolution: 0.02 mm/m,
– vibration frequency < 10 Hz.

With the sensor, the building deflection can be measured accurately, continuously and automatically. The measurement is made by registering the vertical deflection of the laser beam after it passes a liquid-filled wedge (Fig. 3b). The outcome of the measurement, in the form of a graphic printout on a recorder tape, offers a full picture of deflections, both in terms of their value, direction, and duration.

The results of the continuous monitoring of the chimney base deflection are presented in a diagram of changes in the values and direction of the resultant deflection vector (Fig. 4). The monitoring demonstrated that noticeable deflection of the chimney base started on 28 June and lasted until August 2016, i.e. over the period of the strongest impact of the mining activity on the surface of the chimney site. On 8 July, a change was observed in the deflection in the south-eastern direction, and from 14 July - a significant increase in the south-western direction. On 4 and 5 August, the resultant deflection in this direction was 2.3 mm/m and subsequently it gradually diminished and the resultant deflection changed its direction to north-eastern. This tendency was observed until mid-November, when the deflection decreased to 0.72 mm/m. In December 2016, the deflection became stable.

 https://doi.org/10.1051/e3sconf/20183601005

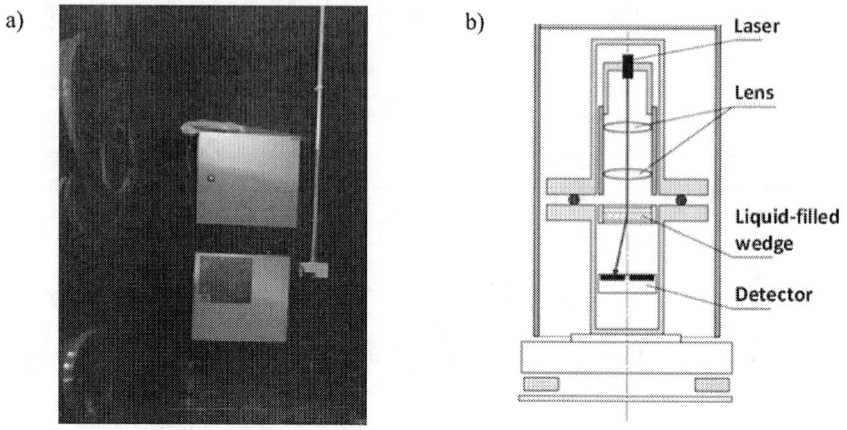

Fig. 3. Laser sensor for continuous measurement of deflections a) a view of the sensor and the recorder, b) the sensor diagram

Compared to the initial position on the date of the commencement of the measurement, the chimney base showed a displacement in the south-western direction, oscillating around a mean value of 1.0 mm/m.

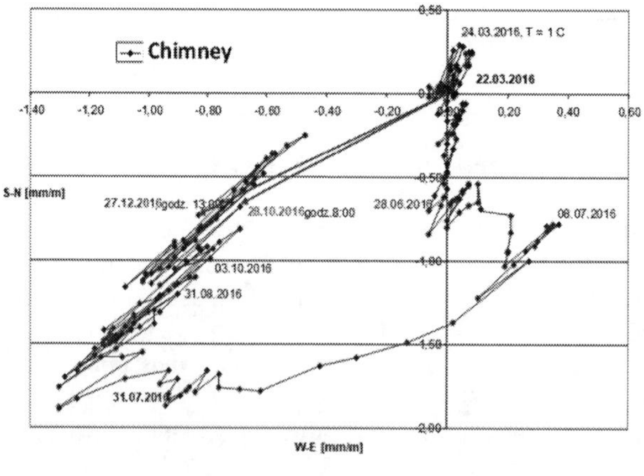

Fig. 4. Chimney base deflection vectors from the date of laser sensor installation

5 Land-based 3D laser scanning

As part of the surveillance, in the period from July 2015 until December 2017, nine measurements of the chimney were performed using ground-based 3D laser scanning. The long-distance 3D laser scanners Trimble TX8 and Riegl VZ-4000 were used.

The 3D Trimble TX8 is an ultra-high speed, phase-shift, long-distance scanner (Fig. 5). It can capture up to one million points per second while maintaining high accuracy over the entire scan range, within the scan range of up to 340 m and broad field of view (317° × 360°).

Riegl VZ-4000 is an impulse terrestrial long-range scanner (Fig. 6). Its operating range is 5 – 4000 m in favourable air conditions (medium air transparency) and low reflectiveness. It works with maximum efficiency within the range of 1 km. Riegl VZ-4000 is among the fastest of its type: it works with a frequency of 50 – 300 kHz (37 – 222 thousand points/s).

 https://doi.org/10.1051/e3sconf/20183601005

Fig. 5. 3D Trimble TX8

Fig. 6. 3D Riegl VZ-4000

The chimney measurements were made from four scanner workstations. Their positioning is shown in Fig. 7 [2]. The data obtained from measurements, in the form of point clouds, was registered. The absolute orientation of the points was determined. The transformation is based on clearly identifiable overlapping points in point clouds. The points can be natural or artificial. Natural targets are the selected chimney points. Artificial targets are spheres inserted in the measured space.

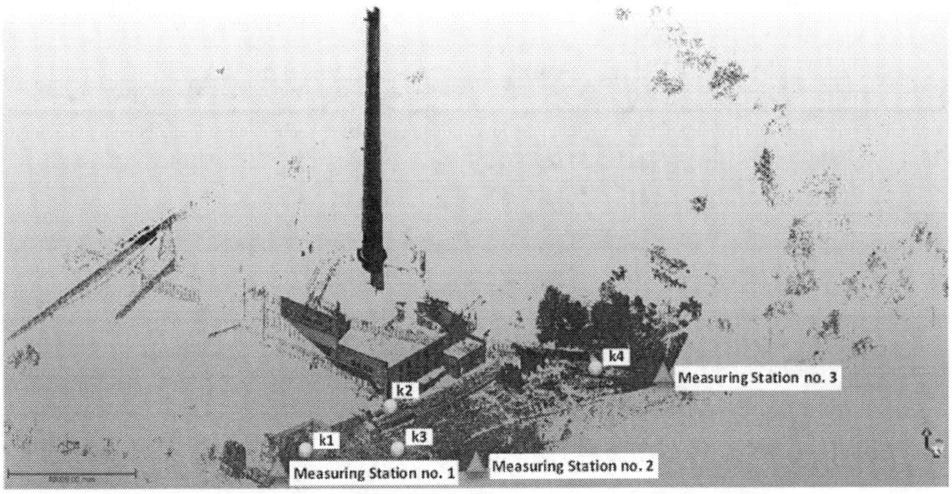

Fig. 7. Layout of measuring stations and spheres (k1-k4) for chimney measurement

The processing of data generated by the 3D laser scanner measurements was done using Trimble Real Works 10.3 software.

The 3D chimney model made from point clouds (Fig. 8) was scrutinised. The chimney cross-sections were prepared, spaced at a distance of 1 m. The location of the circular section centre on successive horizontal lines was analysed, and the direction and value of chimney deflection was determined at each level (Fig. 9).

The measurements are presented in Table 2. Each measurement is accompanied with a brief note on the ambient conditions in which it was taken.

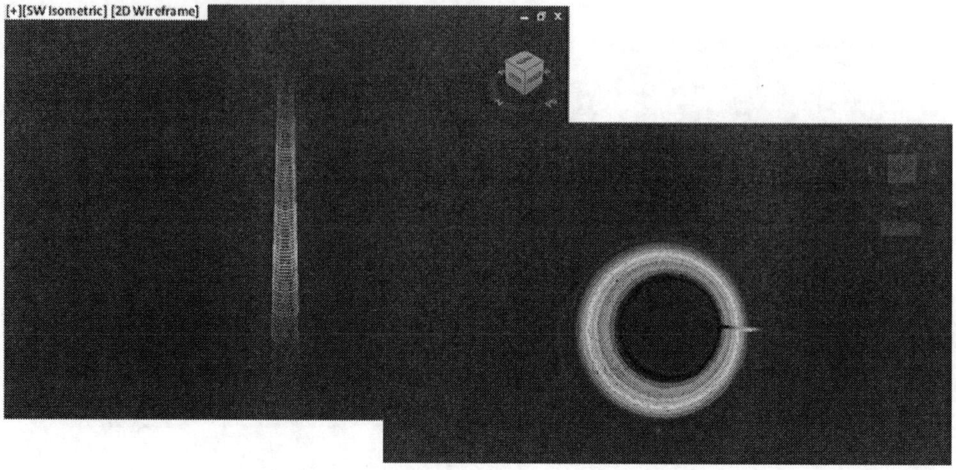

Fig. 8. Chimney model and its cross sections on successive horizontal lines

In December 2017, the chimney deflection was 546 mm in the north-eastern direction (azimuth 62°). With respect to the measurement of June 2017, the deflection decreased by 29 mm, and the azimuth changed by 5° to the southern direction.

Table 2. List of chimney top deflections measured by laser scanning

Item	**Measurement Date**	**Deflection** [mm]	**Deflection Direction** [°]	**Scanner type**	**Ambient conditions** air temperature wind direction and flow velocity
1	29.07.2015	560	55	Trimble TX8	21°C; SW 19 km/h
2	04.11.2015	556	51	Trimble TX8	11°C; WSW 13 km/h
3	18.03.2016	546	54	Trimble TX8	10°C; W 21 km/h
4	21.06.2016	541	64	Riegl VZ-4000	21°C; SW 14 km/h
5	26.09.2016	560	58	Riegl VZ-4000	18°C, EEW 10 km/h
6	18.10.2016	593	58	Riegl VZ-4000	11°C, SSW, 18 km/h
7	22.12.2016	583	60	Riegl VZ-4000	-1°C, SSW, 20 km/h
8	09.06.2017	575	57	Riegl VZ-4000	23°C; SE 9 km/h
9	12.12.2017	546	62	Riegl VZ-4000	10°C; SSW 21 km/h

6 Checks of the technical condition

The chimney was repaired in 1997. Before mining activity started, it showed no major damage, except for some local thermal cracking. During the mining activity carried out within the chimney site, systematic ground-based observation is made using binoculars and cameras equipped with telephoto lens. In October each year, during a several-day technological break in boiler house operation, the condition of the chimney facade and interior is inspected by climbers.

Fig. 9. The drone

In June 2017, a chimney technical check was made using an unmanned aerial vehicle commonly called a drone [3]. The checks were made using a DJI MAVIC Pro drone (Fig. 9).

With the drone, the technical conditions of the external brick wall of the chimney could be assessed, along its entire height, with accuracy comparable to direct visual inspection. The check showed ovalisation of the chimney shaft cross-section at the outlet (Fig. 10) which could not be detected using traditional inspection methods.

In October 2017, during a ground-based inspection of the chimney condition, a crack along a joint was observed at the height of approximately 35 m, from the south-western side. The inspecting personnel indicated that the crack was not clearly visible on previous checks. The analysis of photographs taken earlier indicates that the damage was already there in May 2015. Another inspection of the chimney's technical conditions, made with the drone in November 2017, demonstrated that the technical condition of the chimney shaft wall did not significantly change at this point (Fig. 11).

Fig. 10. Ovalisation of the chimney shaft cross-section at the flue gas outlet level

Fig. 11. Crack at the chimney wall - a photo taken by a drone in November 2017

7 Summary

For decades, observations of buildings and structures in regions impacted by underground mining have been made using two methods, these being geodetic measurements of structural deformations and foundations and visual inspections of the technical condition. In many situations such procedures have proven insufficient. Frequently, their use is prevented by inaccessibility of the object or its part.

The experience and knowledge gained from monitoring the chimney indicates that it is reasonable to use laser sensors which facilitate the continuous measurement of deflection in real time. The measurement results can easily be transmitted, and exceedance of threshold values can be signalled.

With 3D laser scanners, changes in the geometry of buildings and structures can be monitored. Terrestrial scanning is a source of information about the condition of buildings and their damage, caused by mining. The colour 3D image created from millions of pieces of measurement data can be used for further operations and analyses, including measurements and 3D modelling. The data obtained from laser scanning adds a new quality to mining damage inventory and monitoring processes.

References

1. A. Szade : Polish patent no 151 105
2. P. Gruchlik: Naziemne skanowanie laserowe 3D, doświadczenia i perspektywy. Przegląd Górniczy 5. (2015)
3. W. Mika, A. Ferenc, S. Czaja: Monitoring obiektów budowlanych z zastosowaniem drona. Przegląd Górniczy 1(2018)

Approximate classification of mining tremors harmfulness based on free-field and building foundation vibrations

Krystyna Kuzniar[1,*], *Krystyna* Stec[2], and *Tadeusz* Tatara[3]

[1]Pedagogical University of Cracow, Institute of Technology, Krakow, Poland

[2]Central Mining Institute, Department of Geology and Geophysics, Katowice, Poland

[2]Cracow University of Technology, Institute of Structural Mechanics, Krakow, Poland

Abstract. The paper compares the results of an approximate evaluation of mining tremors harmfulness performed on the basis of free-field and simultaneously measured building foundation vibrations. The focus is on the office building located in the Upper Silesian Basin (USB). The empirical Mining Intensity Scale GSI-GZW$_{KW}$-2012 has been applied to classify the harmfulness of the rockbursts. This scale is based on the measurements of free-field vibrations but, for research purposes, it was also used in the cases of building foundation vibrations. The analysis was carried out using the set of 156 pairs ground – foundation of velocity vibration records as well as the set of 156 pairs of acceleration records induced by the same mining tremors.

Keywords: mining tremors, GSI scale, harmfulness of vibrations, ground vibrations, foundation vibrations

1 Introduction

So called paraseismic sources – for example, mining tremors, can induce ground motions. Mining tremors have been observed in many countries all over the world. These phenomena refer to underground exploitation of coal, copper, and gold mines in China, Australia, South Africa, Canada, India, France, Czech Republic, Russia, and Poland [1]. They are random events with respect to their time, place and magnitude, as with earthquakes. The energy of these phenomena is smaller than in earthquakes but can induce negative influence on structures. Mining-induced ground motion is a hazard to surface structures, as ground motions can result in their damage.

In Poland active seismicity and rockbursts are substantial problems e.g. in Upper Silesian Basin (USB) which is the most seismically active mining region in Poland. The mining-induced seismicity hazard in this area has been recently increased. The intensity of

* Corresponding author: krystyna.kuzniar@up.krakow.pl

mining-induced seismic events growing and has stated dangerous for surface structure as well as underground facilities.

Records of mining-induced vibrations (referring to acceleration and velocity) simultaneously registered on the free-field close to a building, and at the foundation level of the building, can differ significantly in this area. It is common property for free-field (ground) and building foundation vibrations in mining areas [1]. It can be observed to maximal values of acceleration and velocity, to response spectra as well as the bands of dominant frequencies. These differences can be explained by the soil-structure interaction effect.

Convenient in practice, simple, approximate method of assessing the harmfulness of vibration of mining buildings is the use of properly prepared scales [2-6]. These scales allow you to determine the degree of vibration intensity and further to estimate their effect on buildings. They are based on soil vibration parameters (e.g. vibration velocity, duration of intense vibration phase). Meanwhile, the vibrations measured on the foundations of buildings may differ significantly from the vibrations recorded on the ground next to the buildings [7-11]. The results presented in [12] indicate that accurate determination of the vibration hazard for buildings is possible on the basis of vibrations of the foundation.

This paper is focused on approximate classification of mining tremors harmfulness based on free-field and simultaneously recorded building foundation vibrations. The research refers to approximate way for assessing the harmfulness of mine-induced vibrations to an actual low-rise office building. To do this, one can use appropriately prepared scales. These scales use the recorded ground or building foundation vibrations and their parameters.

In this study the assessment of the surface vibration intensity according to the GSI-GZW_{KW}-2012 scale was conducted using simultaneously recorded free-field vibration patterns alongside the building and, for cognitive purposes, vibrations of foundations of the building. The influence of the surface vibration measurement site on the degree of vibration intensity was analysed. The Mining Intensity Scale GSI-GZW_{KW}-2012 has been applied in velocity version (the basis scale) as well as acceleration version (supporting version) [13-15]. Additionally, the impact of the vibration measurement site (ground next to the building or foundation of the building) on the compatibility of the degree of intensity according to the GSI-GZW_{KW}-2012 scale in velocity (GSI-GZW_{KW}-2012-V) and acceleration (GSI-GZW_{KW}-2012 –A) versions was investigated.

2 Experimental tests

Mine-induced rockbursts in the Upper Silesian Basin (USB) were the sources of the analysed vibrations. The analysis of the vibration harmfulness is based on the free-field and simultaneously measured building foundation vibrations.

The focus is on the actual masonry office (administrative) building. It is a low-rise (2-storey) building with plan dimensions of 23.8m x 12.5m and 7.30m high. Bearing walls of the building constitute a mixed transverse-longitudinal system. There is no basement in the considered building and it is founded directly on the ground using concrete strip foundations at a depth of 1.4m. The sub-soil consists of a soil layer, medium and fine sand and partially yellow dust.

Acceleration records of free-field vibrations next to the building and on the building foundation were measured simultaneously using so-called "an armed partition" accelerometers. The main attention was devoted to measurements of the horizontal vibration components in the directions parallel to the transverse and longitudinal axis of the building because of the effects of mine-induced rockbursts on a building, as well as the size of building geometric parameters. The records of component vibration velocities were obtained by the acceleration records integration. Resultant accelerations and resultant velocities of free-field and foundation vibrations were computed on the basis of horizontal vibration components in both directions.

Records induced by mining tremors with energies included in the range of $1.0 \cdot 10^5$ J – $4.0 \cdot 10^9$ J and epicentral distances included in the range of 230m – 2045m were taken into account. The analysis was carried out using the set of 156 pairs ground – foundation of velocity vibration records as well as the set of 156 pairs of acceleration records induced by the same mining tremors.

The ranges of the maximum values of the resultant free-field and foundation horizontal accelerations obtained using the components of the recorded horizontal vibrations filtered in the band to 10Hz (PGA_{H10} and PFA_{H10} respectively), as well as the ranges of the maximum values of the resultant free-field and foundation horizontal velocities (PGV_H and PFV_H respectively) are listed in Table 1.

Table 1. The ranges of the maximum values of the resultant free-field and foundation horizontal vibrations: PGA_{H10}, PFA_{H10}, PGV_H, PFV_H.

		Values	
		Minimum	**Maximum**
Accelerations	PGA_{H10} **[m/s²]**	0.01	1.69
	PFA_{H10} **[m/s²]**	0.01	2.29
Velocities	PGV **[m/s]**	0.0004	0.0557
	PFV **[m/s]**	0.0015	0.0909

3 Approximate estimation of the intensity of mine-induced surface vibrations using recorded free-field and building foundation vibrations

3.1 The Mining Intensity Scale GSI-GZW$_{KW}$-2012

GSI-GZW$_{KW}$-2012 scale is an empirical-measurement scale for monitoring and assessing the impact of mining shocks induced in the Upper Silesian Basin (USB) on surface development and perceptibility of vibration by humans [13-15]. It refers to buildings, linear underground infrastructure facilities (underground water, gas, sewage or other networks (e.g. tunnels)). This scale exists in velocity (basis scale) and acceleration (the auxiliary scale) versions, respectively denoted as GSI-GZW$_{KW}$ -2012 –V and GSI-GZW$_{KW}$-2012-A. The GSI-GZW$_{KW}$ -2012 scale associates macro-seismic effects of vibrations with the values of the parameters of surface vibrations, selected for the best description of the actual intensity of the vibrations.

With regard to buildings, the GSI-GZW$_{KW}$-2012 scale allows an approximate assessment of the impact of mining vibrations from harmless vibrations through causing architectural damages, up to the limit above which structural damages may occur. The scale distinguishes five degrees of vibration intensity listed in Table 2. Severity levels are attributed to damage that may occur in buildings. The GSI-GZW$_{KW}$ -2012 scale does not distinguish clearly the technical condition of the building.

The basic parameters necessary in assessing the impact of surface vibrations caused by rockbursts through the use of the scale GSI-GZW$_{KW}$ -2012, are: the maximum value of the resultant of the free-field horizontal velocity PGV_H – scale GSI-GZW$_{KW}$-2012-V, the maximum value of the resultant free-field horizontal acceleration PGA_{H10}, obtained on the basis of the filtered components of the recorded horizontal vibration in the band to 10Hz – scale GSI-GZW$_{KW}$-2012-A, and the duration of the intensive phase of vibrations t_{Hv} and t_{Ha} for the velocity and acceleration respectively [15].

Table 2. Effects of mining shocks referring to degrees of intensity in the GSI-GZW$_{KW}$-2012 scale.

Degree of vibration intensity *I*	Description of effects according to the GSI-GZW$_{KW}$-2012 scale
0	not noticeable; with no effect on buildings
I	being felt; only sporadically causing an increase in existing damage
II	intensification of existing damage; enlargement of pre-existing damage to non-structural elements
III	non-structural damages; first damages of non-structural elements; extensive enlargement of existing damages to non-structural elements
IV	structural damages; single damages of structural elements

3.2 Application of the scale GSI-GZW$_{KW}$-2012 for the evaluation of harmfulness of the analysed vibrations

Graphical illustration of the results of intensity assessment of the free-field vibrations induced by the 156 analysed rockbursts in the cases of GSI-GZW$_{KW}$-2012-V scale as well as GSI-GZW$_{KW}$-2012-A scale applications are shown in Fig. 1 and Fig. 2 respectively. Additionally, for the comparison, analogous results related to the simultaneously measured foundation vibrations are also included in these figures. Next, in Fig. 3 and Fig. 4, the comparisons of the degrees of intensity of free-field and building foundation vibrations determined using the two versions of scales in the cases of the 156 successive rockbursts are presented.

It is visible that some differences can occur in the vibration intensity evaluation depending on the place of simultaneously registration – free-field or building foundation (cf. Fig. 1 – Fig. 4). Besides, the "shift" into the longer durations of the intensive phases of foundation vibrations for the velocities as well as accelerations (t_{Hv} and t_{Ha}) can be observed in the case of intensity assessments based on foundation vibrations in comparison to intensity assessments based on free-field vibrations (cf. Fig. 1, Fig. 2). It should also be noted that the distributions of the maximum values of the resultant velocities as well as accelerations in the cases of the free-field and foundation vibrations are different despite of the same set of rockbursts – the sources of vibrations.

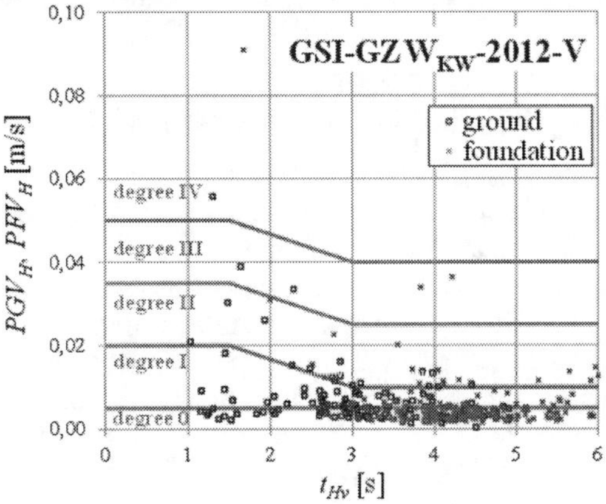

Fig. 1. Results of the assessment of the intensity of surface vibrations induced by the considered mining tremors using the scale GSI-GZW$_{KW}$-2012-V.

https://doi.org/10.1051/e3sconf/20183601006

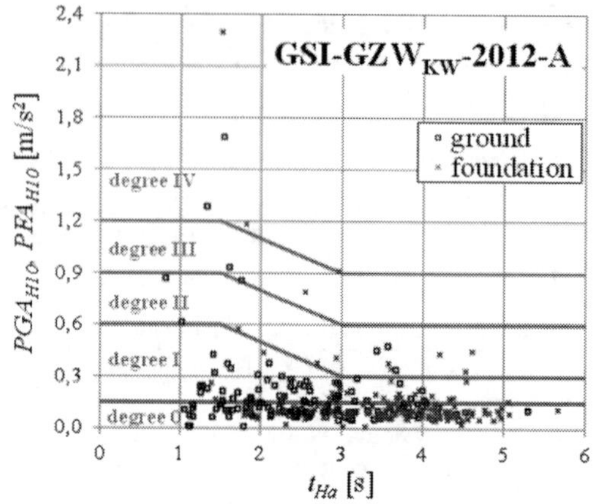

Fig. 2. Results of the assessment of the intensity of surface vibrations induced by the considered mining tremors using the scale GSI-GZW$_{KW}$-2012-A.

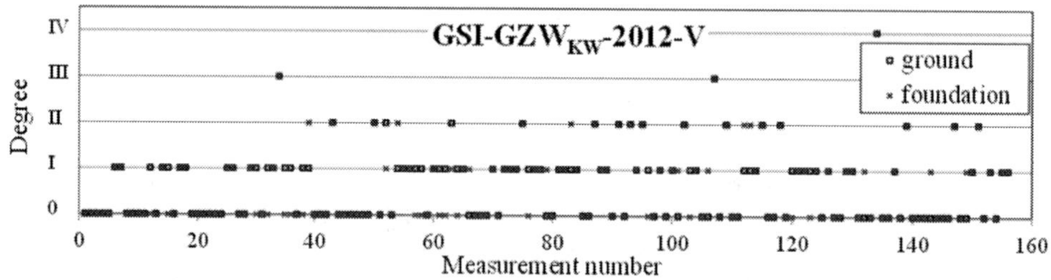

Fig. 3. Comparison of the degrees of intensity of free-field and building foundation vibrations determined using the scale GSI-GZW$_{KW}$-2012-V in the cases of the successive rockbursts.

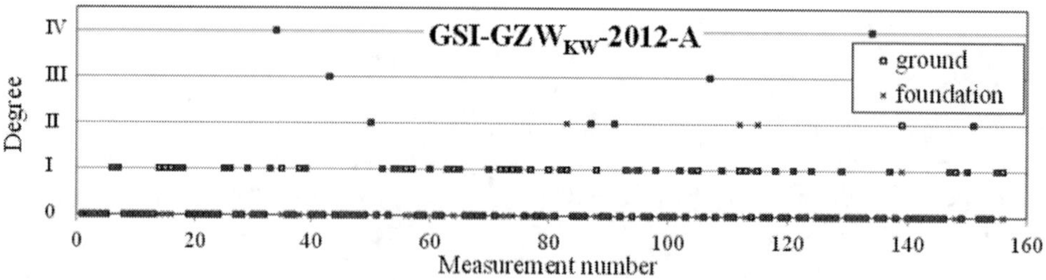

Fig. 4. Comparison of the degrees of intensity of free-field and building foundation vibrations determined using the scale GSI-GZW$_{KW}$-2012-A in the cases of the successive rockbursts.

Table 3 shows the list regarding the rockbursts for which compatibility of the intensity assessments was obtained in the cases of the free-field and foundation vibrations, whereas Table 4 presents the list regarding the mining tremors for which compatibility of the intensity assessments was not obtained in the cases of the ground and foundation vibrations.

Looking at the results collected in Table 3 it can be stated that in the cases of about 80% of all (156) shocks the compatibility of the intensity assessments based on the simultaneously measured free-field and foundation vibrations was obtained using the velocity version of scale GSI-GZW$_{KW}$-2012-V (80.1%), as well as using the acceleration version of scale GSI-GZW$_{KW}$-2012-A (84.6%).

Among the rest considered cases of mine-induced tremors (without the compatibility of the vibration intensity assessments based on the free-field and foundation vibrations) the straight majority of vibration intensity assessments with higher degrees of vibrations intensity for the free-field vibrations than for the foundation vibrations were denoted according both to velocity (61.3%) and acceleration (79.2%) versions of GSI-GZW$_{KW}$-2012 scale (cf. Table 4).

Table 3. List regarding the rockbursts for which compatibility of the intensity assessments was obtained in the cases of the free-field and foundation vibrations.

GSI-GZW$_{KW}$-2012-V			**GSI-GZW$_{KW}$-2012-A**		
Intensity *I*	Number of shocks	Percentage of all (156) shocks	Intensity *I*	Number of shocks	Percentage of all (156) shocks
0	68	43.6%	0	92	58.9%
I	40	25.6%	I	32	20.5%
II	14	9.0%	II	4	2.6%
III	2	1.3%	III	2	1.3%
IV	1	0.6%	IV	2	1.3%
In total	125	80.1%	In total	132	84.6%

Table 4. List regarding the rockbursts for which compatibility of the intensity assessments was not obtained in the cases of the free-field and foundation vibrations.

GSI-GZW$_{KW}$-2012-V			**GSI-GZW$_{KW}$-2012-A**		
Intensity *I*		Number of shocks	Intensity *I*		Number of shocks
free-field	foundation		free-field	foundation	
0	I	7	0	I	2
I	0	17	I	0	18
I	II	5	I	II	3
II	I	2	II	I	1
In total		31 (19.9%)	In total		24 (15.4%)

Table 5. Comparison of the compatibility of the free-field – foundation intensity assessments in the cases of the successive free-field vibration intensity degrees (*IG*).

GSI-GZW$_{KW}$-2012-V			**GSI-GZW$_{KW}$-2012-A**		
IG /Number of shocks	Percentage of the *I* free-field – foundation assessments		*IG* /Number of shocks	Percentage of the *I* free-field – foundation assessments	
	compatible	not compatible		compatible	not compatible
0/75	90.7%	9.3%	0/94	97.9%	2.1%
I/62	64.5%	35.5%	I/53	60.4%	39.6%
II/16	87.5%	12.5%	II/5	80.0%	20.0%
III/2	100.0%	0.0%	III/2	100.0%	0.0%
IV/1	100.0%	0.0%	IV/2	100.0%	0.0%

Comparison of the compatibility of the ground – foundation intensity assessments in the cases of the successive free-field vibration intensity degrees (*IG*) is shown in Table 5. Very fit results of intensity degrees of the free-field and foundation vibrations were determined in the cases of the free-field vibration intensity degrees: *IG* – 0, *IG* – III, *IG* – IV. This conclusion relates to GSI-GZW$_{KW}$-2012-V scale as well as GSI-GZW$_{KW}$-2012-A scale.

But it is worth noting that the number of considered rockbursts with the III degree intensity and the IV degree intensity is not numerous. Hence boldly generalization is not possible in these groups of shocks. Relatively weak compatibility of the vibration intensity assessments based on the free-field and foundation vibrations concerns the first (I) and the second (II) degree of vibration intensity (cf. Table 5).

The influence of the vibration measurement site (ground next to the building or foundation of the building) on the compatibility of the degree of vibration intensity according to the GSI-GZW$_{KW}$-2012 scale in velocity (GSI-GZW$_{KW}$-2012-V) and acceleration (GSI-GZW$_{KW}$-2012 –A) versions is presented in Table 6 and Table 7. It can be stated from the results collected in Table 6 that the similar number of shocks for which compatibility of the intensity assessments was determined in the cases of the GSI-GZW$_{KW}$-2012-V and GSI-GZW$_{KW}$-2012-A scales despite of the vibration measurement site – 117 shocks (75.0% of all) and 119 shocks (76.3% of all) respectively. The most different assessments of vibration intensity were obtained in the cases of the rockbursts classifying to the first and the second degrees of vibration intensity according to the GSI-GZW$_{KW}$-2012-V scale. This conclusion refers both to the free-field vibrations as well as to the foundation vibrations.

Table 6. List regarding the rockbursts for which compatibility of the intensity assessments was obtained in the cases of the GSI-GZW$_{KW}$-2012-V and GSI-GZW$_{KW}$-2012-A scales.

Intensity *I* determined on the basis of:					
free-field vibrations			**foundation vibrations**		
Intensity *I*	Number of shocks	Percentage of all (156) shocks	Intensity *I*	Number of shocks	Percentage of all (156) shocks
0	71	45.5%	0	85	54.5%
I	39	25.0%	I	25	16.0%
II	5	3.2%	II	7	4.5%
III	1	0.6%	III	1	0.6%
IV	1	0.6%	IV	1	0.6%
In total	117	75.0%	In total	119	76.3%

Table 7. List regarding the rockbursts for which compatibility of the intensity assessments was not obtained in the cases of the GSI-GZW$_{KW}$-2012-V and GSI-GZW$_{KW}$-2012-A scales.

Intensity *I* determined on the basis of:					
free-field vibrations			**foundation vibrations**		
Intensity *I*		Number of shocks	Intensity *I*		Number of shocks
GSI-GZW$_{KW}$-2012-V	GSI-GZW$_{KW}$-2012-A		GSI-GZW$_{KW}$-2012-V	GSI-GZW$_{KW}$-2012-A	
0	I	4	I	0	24
I	0	23	II	I	10
II	I	10	II	0	1
II	III	1	II	III	1
III	IV	1	III	IV	1
In total		39 (25.0%)	In total		37 (23.7%)

4 Conclusions

Carried out analyses show that significant differences can occur in the cases of approximate classification (using of GSI-GZW$_{KW}$-2012 scale) of mining tremors harmfulness based on the free-field vibrations and simultaneously measured building foundation vibrations. The most of incompatibilities in assessments of vibration intensity (degree of intensity) are observed in the cases of the first and the second degree of vibration intensity. Whereas the good compatibility of classification of vibration intensity on the basis of free-field or foundation vibrations is visible in the cases of very weak rockbursts (with vibration intensity of degree 0) and, on the other hand, in the cases of very strong shocks (with vibration intensity of degree III and IV). However, because of the small number of vibrations with high degree of intensity (III and IV degree) in the set of considered mine-induced vibrations, the conclusions regarding such vibrations require verification in the future investigations.

References

1. E. Maciag, K. Kuźniar, T. Tatara, Earthquake Spectra 32, 3, 1769–1791 (2016)
2. A. Biessikirski, Przegląd Górniczy 6, 1–6 (2014)
3. J. Dubiński, K. Jaśkiewicz, A. Lurka, G. Mutke, Górnicza skala intensywności sejsmicznej GSI-2004/11 dla wstrząsów górniczych w LGOM. [in:] Weryfikacja skal GSI-2004, Wrocław: KGHM CUPRUM sp. z o.o. – CBR, 150–169 (2011) (unpublished work)
4. J. Dubiński et al., Weryfikacja skal GSI-2004, KGHM CUPRUM sp. z o.o. – CBR, (Wrocław 2011) (unpublished work)
5. I. Jaśkiewicz-Proć, Przegląd Górniczy 6, 14–23 (2014)
6. K. Kuźniar, T. Tatara, Procedia Engineering 199, 2378–2383 (2017)
7. K. Kuźniar, E. Maciąg, T. Tatara, First European Conference on Earthquake Engineering and Seismology (ECEES), Geneva, Switzerland, Abstract Book, 466–467 (2006) (full paper on CD)
8. K. Kuźniar, T. Tatara, Przegląd Górniczy 10, 31–36 (2015)
9. E. Maciąg , Inżynieria Morska i Geotechnika 4, 240–250 (2006)
10. E. Maciąg, K. Kuźniar, T. Tatara, [in:] Stypuła K. (ed.), Aktualne problemy wpływów sejsmicznych i parasejsmicznych na budowle. Badania wstrząsów górniczych i drgań komunikacyjnych, 477/2, Seria Inżynieria Lądowa, 39–66 (Wyd. PK, Kraków, 2015)
11. T. Tatara, Odporność dynamiczna obiektów budowlanych w warunkach wstrząsów górniczych (Wydawnictwo PK, Kraków 2012)
12. E. Maciąg, Inżynieria i Budownictwo 12, 670–677 (2005)
13. A. Barański, L. Kloc, T. Kowal, G. Mutke, Bezpieczeństwo Pracy i Ochrona Środowiska w Górnictwie 6, 3–10 (2014)
14. A. Barański, L. Kloc, T. Kowal, G. Mutke, Bezpieczeństwo Pracy i Ochrona Środowiska w Górnictwie 11, 32–33 (2014)
15. J. Dubiński, G. Mutke, T. Tatara, L. Muszyński, A. Barański, T. Kowal, Zasady stosowania zweryfikowanej Górniczej Skali Intensywności drgań GSI-GZW$_{KW}$-2012 do prognozy i oceny skutków wstrząsów indukowanych eksploatacją złóż węgla kamiennego w zakładach górniczych Kompanii Węglowej S.A. na obiekty budowlane i na ludzi, Kompania Węglowa, Katowice (2013) (unpublished work)

Reliability and safety, and the risk of construction damage in mining areas

Izabela Skrzypczak[1*], *Janusz* P. Kogut[2], *Wanda* Kokoszka[1], and *Grzegorz* Oleniacz[1]

[1]Rzeszow University of Technology, Faculty of Civil, Environmental Engineering and Architecture, Rzeszow, Poland

[2]Cracow University of Technology, Faculty of Civil Engineering, Krakow, Poland

Abstract. This article concerns the reliability and safety of building structures in mining areas, with a particular emphasis on the quantitative risk analysis of buildings. The issues of threat assessment and risk estimation, in the design of facilities in mining exploitation areas, are presented here, indicating the difficulties and ambiguities associated with their quantification and quantitative analysis. This article presents the concept of quantitative risk assessment of the impact of mining exploitation, in accordance with ISO 13824 [1]. The risk analysis is illustrated through an example of a construction located within an area affected by mining exploitation.

Keywords: risk analysis, safety, damage, mining area

1 Introduction

Reliability, safety, and analysis of the risk of damage to building structures has at least two meanings in the literature: in general, qualitative and quantitative. General reliability is defined as the ability to perform the planned construction of a function at a certain time of use or as a specific condition of the structure. In particular, reliability is the probability that the designed structure does not exceed the limits specified in the planned operational time. Certain measures are taken to ensure the reliability of a structure. They include, in addition to relevant calculations, a range of activities related to ensuring durability and quality. These measures also aim to reduce calculation errors, as well as operational errors and the serviceability of structures [2]. Therefore, reliability is associated not only with the exceeding of conventional limit states, but also with the elimination and/or reduction of other hazards (e.g., in the case of construction works in mining areas with the limitation of the predicted impact of mining operations on the technical condition of buildings). While safety is narrower concept of reliability, it is usually associated with a lack of threat and often identified with the complement of the risk of human health, as well as with the economic, social, and environmental losses in the projected lifetime. In this narrower sense,

* Corresponding author: izas@prz.edu.pl

safety is also referred to as the conditional reliability. Safety is defined as the probability that a structure, that meets the relevant requirements and conditions, will not reach certain arbitrary limit states during the assumed lifespan. The objects are designed for mining areas so that they are reliable and safe during their use, but they must be properly built and serviced. The exploitation of mining areas initiates a slow movement of rock masses and soil towards the excavation. The influence of exploitation on the surface of the site manifests itself in the form of uneven depressions, terrain deformations, and/or unfavorable changes in the level of the water table. Buildings are subject to various stress and strain conditions as a result of variable interactions. Due to terrain deformation, rigid (low deformation) structures will experience additional loads, while flexible (deformable) structures experience a more complex state of deformation. Therefore, construction objects in mining areas, in order to be reliable and safe to use, must be properly designed and constructed to predict the impact of mining operations. Knowledge of these problems concerns not only designers, but also contractors, supervisors, and those persons responsible for quality control and compliance tests in construction companies. Unfortunately, without an understanding of the assumptions and foundations of the theory of reliability, the application may lead to serious errors and misunderstandings in the design of building structures in mining areas. The detailed requirements regarding reliability are specified in the design standards of construction and in safety requirements in construction law, while the risk management requirements are formulated in quality and reliability standards.

2 Reliability management

In accordance with the standards for the design of building structures [3], the required reliability of building structures should be ensured by: designing and properly performing and undertaking quality assurance measures at every stage of construction project implementation. Various levels of reliability can be adopted to ensure ultimate and serviceability limit states of the structure. When choosing levels of reliability for a structure, it is recommended to take into account several relevant factors, including: possible causes of limit states, possible consequences of damage (such as life and health hazards), potential material losses, social reactions to existing damages, costs and procedures, and necessary procedures to be conducted in order to reduce the risk of damage.

The reliability of a structure can be determined by the classification structure as a whole or by the classification of its components. Reliability, relating to the bearing capacity and serviceability, can be achieved by the appropriate combination of: - preventive or protective measures; - calculation means (i.e. representative values of impacts and selection of partial coefficients); - actions and measures regarding quality assurance; - measures taken to reduce calculation, construction, and human errors; - measures concerning the following issues: basic requirements, degree of compactness, durability (including design lifetime), scope and quality of preliminary soil investigations and arrangements for possible impacts of mining operations, accuracy of accepted calculation models, construction details, proper implementation, adequate inspection, and maintenance in accordance with the procedures given in the project documentation.

Measures taken to prevent accidents and limit their consequences can be considered as interchangeable, to a narrow extent, provided that the required levels of reliability are maintained. Procedures and measures of the reliability management are formulated in order to enable the differentiation of the reliability of the structure and its elements (compare to Annex B [3]).

 https://doi.org/10.1051/e3sconf/20183601007

3 Strategies for designing buildings

European norm [4] distinguished two basic strategies for structure design to work against the accidental actions in exceptional computational cases: a) strategies based on accidental actions and values established on the basis of statistical research or contractual values and based on design of the structure to bear accidental loading; structural design for sufficient minimum load-bearing; preventing or reducing the occurrence of accidental actions, b) strategies based on a limited range of local losses in the case when accidental impact is undefined or has a very large variation or frequency of occurrence is difficult to assess, consisted of: stiffening of structure; designing key elements for accidental impact of a normative value; compliance with standard conditions defined in the relevant codes, etc. As far as the design of buildings is concerned, due to the limitation of the allowable level of local destruction of undetermined reasons, there are four Consequence Classes defined depending on the type of building and its use, with the recommended number of strategies to ensure a satisfactory level of structural resistance in order to sustain localized failure without a disproportionate level of collapse: a) Class 1 - with no specific consideration necessary with regard to accidental actions, when a building is designed and constructed in accordance to the rules given in Eurocodes; b) Class 2a (Lower Risk Group) - as a complementary recommendation to Class 1 with providing of effective horizontal ties or effective anchorage of suspended floors to walls respectively from framed and load-bearing wall construction; c) Class 2b (Upper Risk Group) - provided with effective horizontal ties for the framed and load-bearing wall construction, respectively, together with effective vertical ties in all supporting columns and walls, or to ensure that upon any casual removal of each supporting pillar and each beam supporting a column, or any nominal section of load-bearing wall (one at a time in each storey of the building) that the building remains stable and that any local damage does not exceed a given limit; d) Class 3 - systematic risk analysis of the building should be undertaken, taking into account both the normal hazards that may reasonably be foreseen and all abnormal hazards. Strategies recommended for the design of a building with Consequences Classes 1 and 2a are based on standard procedures and have already been applied in practice. In the case of buildings belonging to the Class 2b, verification against advanced numerical analysis is additionally required. In relation to the Class 3, dealing with buildings with frequently applied practical design (including buildings where a great many people have access), it is required to carry out systematic risk analysis that takes into account normal (predictable) and abnormal (unpredictable) hazards. However, the guidelines for the analysis and risk assessment offered by these standards, particularly the quantitative one, are rather general and their application, in practice, raises many doubts, requires specialized knowledge, is hard to reach, and often provides incomplete and inaccurate data. In order to achieve adequate structural stability and resistance disproportionate to cause and progressive destruction, it is recommended to use binding systems and wreaths, to diversify the directions of support of the horizontal floor slabs and to ensure that the conditions of the ties or membrane function after the destruction of their supports, and the partition of construction, etc. [2].

4 Procedures of the building's risk assessment in mining areas

The basis for assessing the risk of mining damage in building facilities is the forecast of the impact of the planned mining operation. The traffic plan should indicate the method used to calculate the deformation rates and should give the assumed values for parameters and coefficients. The assumptions for the forecast should include the results of surface deformation measurements [5]. It should be emphasized that neither the applicable legal provisions nor the method of its execution or the qualifications of contractors define the scope of the inventory of buildings. The GIG instruction [6], recommended for use in the Polish underground mining by the Commission for Surface Protection at the Higher Mining Office, distinguishes two procedures for assessing the resistance of buildings to permanent surface deformation. In the case of buildings with a masonry (longwall) load-bearing structure, from small-sized elements designed without taking into account the impact of mining operations, possibly protected during their use, it refers to the study [7]. As for those buildings with a different construction and buildings subject to special protection, it indicates the need to conduct an assessment. The potential damage of building's evaluation, as a result of mining subsidence, may be assessed by: the expected surface deformations, structural characteristics of construction, properties of building materials and static features of the building system, mechanical properties of the substrate, and current technical condition of the structure. The basic criteria determining the resistance of a building object to a mining subsidence is to check the bearing capacity and serviceability limit state. The method of assessing the safety of the structure results from the adopted method of assessing the resistance [8]. The resistance to the horizontal deformations ε of the surface and the curvature K is usually assessed. These surface deformation factors determine the stress state in the structural members. Resistance of buildings is not a permanent feature and may change due to technical conditions and the influence of the mining subsidence [9,10]. The assessment of the resistance of buildings to a continuous surface deformation is based on the determination of the permissible values of K and ε, while maintaining the safety of buildings, causing a slight nuisance of their use due to cracks of width d and shear strains γ_k [6]. The resilience of existing buildings to mining subsidence depends on their structural characteristics, static features of structure, soil characteristics, and current technical condition [11]. The resilience of buildings designed and constructed for mining impacts is the resistance resulting from the values of curvatures K and horizontal strains ε adopted in projects. This depends on the correctness of the constructional protection made, the current technical condition of the buildings, and the impact on the mining operations. The resilience of buildings to continuous deformation of the substrate may be assessed on the basis of a detailed analysis of their structure, approximate methods, and expert knowledge [8]. The approximate methods have been developed for buildings with a masonry bearing structure. The commonly used methods to assess the resistance of buildings include: a detailed method, based on the diagnosis of the technical condition of the structure, taking into account the predicted impacts of mining exploitation [8]; a point method [12,13]; a modified point method [14,15]; - a resistance scale method [16]; a parametric method [17]; - a safe estimation method [18]; - expert knowledge [11]. Currently, the modified point method (i.e., a version developed by GIG) is commonly used [14,15]. The modified point method, like other earlier versions of this method, consists of assigning an appropriate number of points for each of the seven distinct features of the building and its subgrade. The modified point method is being developed together with detailed instructions for its application, and it allows for the assessment of the resistance of both free-standing and multi-segment buildings. Kawulok [8] established the detailed rules for assessing the resistance of buildings to the impact of mining surface deformations. There are four groups of objects: buildings with a longitudinal load bearing structure, mainly residential and public buildings; other buildings with a rigid structural design, with different solutions of the support system (e.g., skeletal buildings); buildings of various uses, including residential and public buildings; industrial objects such as industrial halls, storage facilities, and other purpose objects which are not belonging to the groups formerly mentioned; engineering structures and structures of special design or other types, such as churches or worship objects, or those requiring special protection (e.g., historic buildings). The safe estimation method is associated with a preliminary assessment of the hazards of buildings influenced by the expected mining operation. This method gives more reliable results than the method of comparing building resistance categories with mining area categories [18,19]. The application of the safe estimation method requires determining, among others, the difference between limit resistance and critical resistance, and it creates certain difficulties [18,19]. For this reason, the division of buildings into safe and unsafe (according to safe estimation) and an assessment of the expected degree of damage should be supplemented by information on the probability of damage and risk to buildings subject to damage in mining areas [18].

5 Risk analysis

The mining exploitation in urban areas, as well as any other economic activity, is associated with risk and uncertainty. The risk may be defined as the product of probability and negative consequences of occurrence of adverse events. The threat is the possibility of an event causing the loss of life, health, and/or social, as well as physical and ecological losses. Hazard identification and identification of possible disaster scenarios or design failure and the assessment of their probability and the consequences that may result belong to the fundamental tasks related to risk analysis. The risk associated with the different stages of the construction process varies and depends on each stage of the threats. Due to the causes of accidents, two types of threats can be distinguished: natural and anthropogenic. Natural hazards are associated with the accidental nature of actions on buildings, as well as random properties of materials and the geometrical dimensions of the structural elements from which they are made. The threat of an anthropogenic, human factor is directly related to the construction process and results from unintended or deliberate derogate from the principles and rules of construction activities (errors and omissions made by human beings). In civil engineering, the risk is a measure of danger, defined as the combination of probability and consequences of the occurrence of an undesirable event. [20,21]. In the case of the random nature of events, they are treated as random events and the risk is a determinate or random value that can be calculated using the following formula:

$$R = \sum_{i=1}^{n} P_i \cdot S_i \tag{1}$$

where: P_i – the probabil0ity of damage to a structure in the area of mining activity, S_i – effects related to the damage of the structure in the area of mining activity. Thus, the objective risk in the design of structures in mining areas is related to the reliability of the structure.

6 Risk quantification

EN 1991-1-7: 2006 norm [4] provides two methods of risk analysis for buildings and structures: the qualitative risk as a part of risk analysis, which identifies all hazards and corresponding hazard scenarios. It requires a detailed examination and understanding of the system. It also requires that the use of buildings and its implications for safety of use is acceptable. For such a reason, a number of techniques of risk analysis have been developed to assist e.g., Process Hazard Analysis (or evaluation) - PHA, Hazard and Operability study - HAZOP [22], fault tree, event tree, decision tree, casual networks, etc. [23,24,25], the quantitative risk as a part of risk analysis probabilities estimated for all undesired events and their subsequent consequences, as it stands in ISO 13824 [1], it may be given by the following formula:

$$R = \sum_{i=1}^{n_H} p(H_i) \sum_{j=1}^{n_p} \sum_{k=1}^{n_s} p(D_j \mid H_i) p(S_k \mid D_j) C(S_k) \tag{2}$$

where, it is assumed that the structure is subjected to n_H different hazards, that may cause damage to the structure in n_p different ways, where the complex state of a damaged structure may be considered in n_s unfavorable conditions of S_k resulting in corresponding consequences $C(S_k)$, $p(H_i)$ is the probability of the occurrence of the *i*-th hazard, $p(D_j|H_i)$ is the conditional probability of the j-th state of causing structural damage to the i-th given hazard, and $p(S_k|D_j)$ is the conditional probability of the *k*-th adverse overall structural performance S_k given the *j*-th damage state. The quantitative analysis of risk calculated as the value of risk in monetary units according to formula (1) should be treated as nominal size, which has no direct reference to the financial outlay incurred in the event of structural damage. Considering that the maximum tolerable probability of structural damage p_{fd} in the full life cycle cost is equal to $C(S)$, it qualifies for the relevant class of reliability ($RCX = \{RC3, RC2, RC1\}$). For the reference period of T_0, as defined in EN 1990: 2012. Basis of structural design as a measure of risk associated with the analysed exceptional situation, one may take a risk index i_R :

$$i_R = \frac{R}{R_{ac}}; R_{ac} = p_{fd}(RCX; T_0) \cdot C(S) \tag{3}$$

Common quantitative risk analysis covers: an estimation of the probability of the possible risks for fixed intensity; an estimation of the probability of various failures and their consequences for the considered threats; an estimation of the probability of adverse reaction to local structural damage and the consequences related to it. The analysis of the causes of damages to structures indicates that they were scarcely associated with insufficient values of parameters adopted in the construction documentation, and their increase would not have prevented the damage [26]. Usually, the reasons for accidents are such factors or circumstances that are not contained in the design process and/or an evaluation of the technical conditions. Hence, a clear analysis and risk assessment is recommended as the most appropriate and promising method to ensure a satisfactory level of resistance to accidental loads and other impacts acting on the structure, including the impacts from mining operations.

7 Risk assessment

The key issue of controversy arousing in the risk assessment is to determine the risk acceptance criterion. There are numerous proposals for quality criteria and mixed criteria, which are usually not very precise, leading to substantially different results. Most often, these are different variants of the principle of ALARP (i.e., the risk as low as reasonably practicable). European norm [4] presented a mixed criterion as the mathematical expectation of the consequences of an undesired event – see Fig. 1.

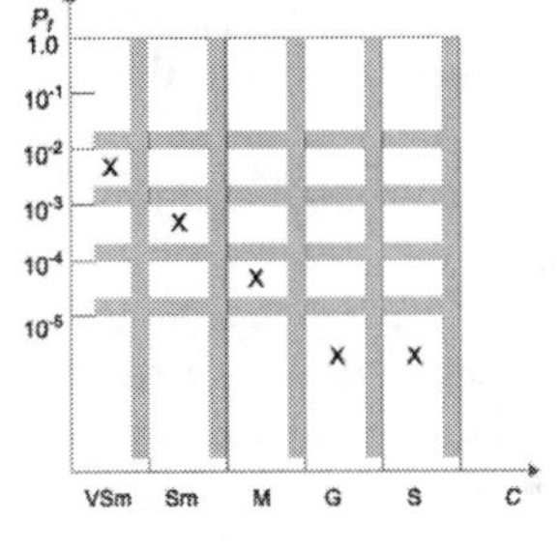

X - represents the largest acceptable level of risk; and **C** - Consequence: **VSm** -Very Small, **Sm** - Small, **M** - Moderate, **G** - Great, **S** – Serious, and $\mathbf{P_f}$ – Probability

Fig. 1. Consequences diagram [4].

8 Risk analysis in mining area – case study

The analysis is performed to a building classified to the consequence Class 3, with monolithic structure of reinforced concrete columns and slabs built in the mining area. Accidental action on the structure is caused by the occurrence of a hazardous landslide and is marked as H_l, (e.g. the threat of mining influences through the occurrence of horizontal deformations and sudden depression of the area), which occurs with probability of $p(H_l)$ = 0.01, and may cause either the local damage to the building D_l or destruction of the foundation D_2 with the conditional probability equal to: $p(D_l|H_l) = 0.1$ and $p(D_2|H_l) = 0.01$. The probability $p(H_l)$ is determined by taking into account the specificities of the mining area (i.e. high inconvenience of using a building due to the influence of continuous surface deformation), hence the value of 0.01. The effect of local defects in both structures, is defined as the damaged part of the structure S_l (up to 100 m^2 floor or the floor area of 15% (EN 1991-1-7: 2006)) or destruction of larger parts or the entire structure S_2, where the conditional probabilities of occurrence are as follows: $p(S_l|D_l) = 0.1$; $p(S_2|D_l) = 0.01$; $p(S_l|D_2) = 0.5$; $p(S_2|D_2) = 0.05$. The consequences of partial damage to the building and the whole destruction are estimated, respectively: $C(S_l)$ = €750 000 and $C(S_2)$ = €15.000 000. On the basis of formula (1), the risk value is equal to $R = 833.25$. An acceptable risk of damage to the structure is deemed to have a reliability class *RC3* for the reference period and is equal to $T_0 = 50$ years, and by taking into account the cost of investment in the full life cycle of $C(S)$ = €7.500 000, is equal to: $R_{ac} = 8.5 \cdot 10^{-6} \cdot 7.5 \cdot 10^{6} = 63.75$. The ratio of risk associated with the destruction of the structure, as a result of the present emergency situation and tolerable risk is: $i_R = R/R_{ac} = 833.25/63.75 = 13.07$. This means that it exceeds the acceptable level of risk by more than 13 times. It is, therefore, necessary to take appropriate action to reduce it; for example, by applying a more effective procedure related to economic evaluation of continuous deformation effects on the building resistance, saving the building from the effects of mining exploitation through appropriate protection of the structure and its foundations, and taking into account the related costs of protection.

9 Final remarks

The assessment of damage to buildings in mining areas is an inherent part of spatial and financial management. Such an analysis is carried out for the planned project (e.g., the further exploitation of a given coal deck), which may endanger the safety of people and structures and the future investment in mining areas. The optimal criterion for designing, dimensioning, and assessing the condition of structures in mining areas allows for the inclusion of quantitative and qualitative requirements, and it is the criterion of minimum safety. The current recommendations for quantitative analysis of risk assessment are quite general and have raised a number of concerns. They are mostly due to the interpretation and quantification of the probability of occurrence of threats, their local and global effects, and the risk consequences. The risk assessment in structural design involves determining the class of structural reliability (acceptable likelihood of destruction), the number of potential casualties, and either the financial, social, ecological, or other consequences. It is particularly difficult to assess the impact of structural damage - the cost of loss. In the case of extractive operations, it is important to be aware of the relationship between loss costs and prevention costs [7]. The total cost of mining prevention, building prevention, and disposal is dependent on priorities. If surface protection is preferred, then construction investment and disposal expenditures may be higher than mining investment expenditures. In some cases, this may lead to the abandonment of mining. If intensive exploitation is preferred, the cost of mining protection rises. At the same time, the degree of nuisance of the influence of exploitation on surface development increases. It is then necessary to increase the cost of mining prevention, in relation to the costs of building prevention and the removal of the effects of mining exploitation, to allow for a greater deformation of the surface or a greater intensity of mining shocks. Often, the cost analysis does not take into account factors such as company image, public trust, and the influence of public opinion. These costs are not measurable, but they may affect business, credibility, business relationships, etc.

References

1. ISO 13824: 2009. Bases for design of structures - General principles on risk assessment of systems involving structures. (2009). International Organization for Standardization
2. Wolinski, Sz. (2006), Risk reliability-based design, in Proceedings of the 11th International Conference on Metal Structures, London-Rzeszów: Taylor & Francis, pp. 887–895
3. EN 1990: 2012. Basis of structural design
4. EN 1991-1-7: 2006. Accidental actions Part 1-7: General actions - Accidental actions
5. Wytyczne w zakresie minimalnych wymogów treści planów ruchu podziemnych zakładów górniczych w aspekcie ochrony powierzchni, Wyższy Urząd Górniczy, Katowice 2013
6. Instrukcja nr 12, Zasady oceny możliwości prowadzenia podziemnej eksploatacji górniczej z uwagi na ochronę obiektów budowlanych, Wyd. GIG, Katowice 2000
7. Kwiatek J. (ed.), (1997) Ochrona obiektów budowlanych na terenach górniczych. Katowice, Wydawnictwo GIG
8. Kawulok M. (2007), Kryteria oceny odporności obiektów budowlanych na ciągłe wpływy eksploatacji górniczej. Prace Naukowe GIG: Górnictwo i Środowisko, V/2007
9. Kwiatek J. (2007), Obiekty budowlane na terenach górniczych, Katowice, Wyd. GIG
10. Kwiatek J. (2010), Ocena niezawodności budynków na terenach wstrząsów górniczych, Górnictwo i Geologia, z. 2, (t. 5), pp. 121-131
11. Mika W., Kaszowska O. (2015), Kryteria dopuszczania eksploatacji górniczej pod terenami zabudowanymi, Przegląd Górniczy, pp. 44-49
12. Przybyła H., Świądrowski W. (1968), Określenia odporności istniejących obiektów budownictwa powierzchniowego na wpływ eksploatacji górniczej, Ochrona Terenów Górniczych, No. 4
13. Lejczak W., Kiczan W., Lubas B., Stranz B., and Bojarski Z. (1969): Zasady stosowania budownictwa zastępczego na terenach górniczych. Wyd. „Śląsk", Katowice
14. Gil-Kleczeńska B., Mika W., and Soczawa Ł. (1990): Zmodyfikowana metoda punktowa oceny odporności obiektów budowlanych zabudowy mieszkalnej na wpływy eksploatacji górniczej o charakterze ciągłym. Materiały V Konferencji „Budownictwo na Terenach Górniczych". Katowice-Kamień k. Rybnika
15. Mika W. (2006), Zmodyfikowana metoda punktowa oceny odporności budynków w świetle dotychczasowych badań i doświadczeń. Prace Naukowe GIG: Górnictwo i Środowisko
16. Kawulok M., Lipski Z. (1995), Podstawy statystycznej metody oceny odporności budynków na wpływy eksploatacji górniczej, Materiały KN-T, III Dni Miernictwa Górniczego i Ochrony Terenów Górniczych, Ustroń - Zawodzie, pp. 305-309
17. Fedorowicz J., Kawulok M. (1997), Odkształceniowe kryterium oceny odporności budynków na terenach górniczych, Prace Naukowe GIG, Seria Konferencje, Katowice, pp. 33-47

18. Ostrowski J. (2006), Deformacje powierzchni a zagrożenie uszkodzeniami budynków na terenach górniczych w ujęciu probabilistycznym, Rozprawy monografie 160, Uczelniane Wydawnictwa Naukowo-Dydaktyczne, Kraków
19. Ostrowski J., Ćmiel A. (2008), The use of a logit model to predict the probability of damage to bullding structures in mining terrains, Archives of Mining Sciences, 53 (2), pp. 161-182
20. JCSS Probabilistic Model Code Part 3: Resistance Models, 10.10.2000
21. Pilecka E. (2007), Statistical analysis of the relation between locations of high energy epicenter tremors and lineaments in areas of the Upper Silesian Basin, Mineral Resources Management, 23 (4), 101-109
22. Steward, M., Melchers, R. (1997). Probabilistic risk assessment of engineering systems. London: Chapman Hall
23. Baker, J., Schubert, M., and Faber, M. (2008). On the assessment of robustness. Journal of Structural Safety, 30 (3), pp. 253-267
24. Faber, M. (ed.), (2008). Risk Assessment in Engineering Principles. System Representation and Risk Criteria. Zurich: Joint Committee on Structural Safety.
25. Knoll, F., Vogel, T. (2009). Design for robustness. Structural Engineering Documents, 11, pp. 253-267
26. Matousek, M. (1977). Outcomings of a Survey on 800 Construction Failures. In IABSE Colloquium on Inspection and Quality Control. ETH Zurich: Institute of Structural Engineering
27. Popiołek E. (1989), Ochrona terenów górniczych, Skrypt nr 1172, AGH Kraków

Some aspects of forecasting the post-mining substratum deformation for evaluation of its influence on constructions

Piotr Strzałkowski[1], *Roman* Ścigała[1], *Katarzyna* Szafulera[1*]

[1] Silesian University of Technology, Faculty of Mining and Geology, Gliwice, Poland

Abstract. Some problems have been discussed, connected with performing predictions of post-mining terrain deformations. Especially problems occur with the summation of horizontal strain over long time intervals as well as predictions of linear discontinuous deformations. Of great importance in recent years is the problem of taking into account transient values of deformations associated with the development of extraction field. The exemplary analysis has been presented of planned extraction influences on two characteristic locations of building structure. The proposal has been shown of calculations with using transient deformation model allowing to describe the influence of extraction advance influence on the value of coefficient of extraction rate *c* (time factor), according to own original empirical formula.

Keywords: post-mining deformations, transient deformation state

1 Introduction

For the first time, mining operations were carried out under the built-up area at the beginning of the XIXth century. Mining works under the Belgian city of Liege caused damage to buildings, which prompted local mining engineers to conduct geodetic surveys. The results of these measurements became the basis for the development of the first theories of forecasting the impact of underground mining operations on the surface. The first solutions were often characterized by low accuracy or were mathematically incorrect. Conducted for many years systematic research related to the above topics allowed to develop a number of theories that are characterized by sufficient accuracy of prediction for mining practice. The review of the theories of mining area deformation, starting with the first methods, can be found among others in [1, 2, 3, 4, 5].

The theory of W. Budryk - S. Knothe [6] enjoys the greatest popularity in Poland. It is characterized by the simplicity of the mathematical model. Long period of its use has resulted in a very good understanding of the parameters that should be taken into account for given mining - geological conditions. An additional reason for its popularity is the fact

* Corresponding author: katarzyna.szafulera@polsl.pl

that a number of software have been developed that allow easy and quick calculations [7, 8].

Considering the above statements, one could mistakenly believe that performing forecasts of the impact of mining operations on the surface is a very simple task. Despite many years of experience in this area however, there are still some doubts and questions for which there are no unambiguous answers.

The first of the doubts is connected with the summation of the values of deformation indices in long time intervals. According to S. Knothe [6], horizontal strain should be summed up in 10-year periods. B. Skinderowicz presented the view, that the values of deformations indices should be reduced by half in two-year periods. B. Drzężla [9], J. Kwiatek [10, 11] and J. Białek [7] considered the concept of the so-called effective deformations, that bases on the assumptions taken from rheological models. J.Białek incorporated in his software a functional relationship, that expresses the reduction of strain caused by the extraction of a given mining field for the purpose of assessment of its influence on the building object:

$$\varepsilon(t)_e = \int_0^t \frac{d\varepsilon(\tau)}{d\tau}\left[A_{rel} + (1 - A_{rel})e^{-(t-\tau)/T_{rel}}\right]d\tau \tag{1}$$

where: $\varepsilon(t)$ - actual horizontal strain [‰],
$\varepsilon(t)_e$ - effective horizontal strain [‰],
t - time [years],
A_{rel} - relaxation coefficient,
T_{rel} - relaxation time [years].

Fig. 1 graphically presents the mentioned above strain reduction methods, according to the works [4, 12].

In the Upper Silesia Basin, intensive mining extraction has been carried out for over 200 years. If we were to sum up the values of horizontal strain in some regions for long periods of time, it would reach a very large value. When considering the possibility of transferring such large deformations from subsoil layers to given construction, one would have to conclude that carrying out such extraction would lead to mass construction disasters which - in fact, did not take place, of course. In the light of this statement, so-called phenomena of "strain relaxation" in accordance with formula (1) seems to be a reasonable solution. An alternative to this proposal may be to assess the objects resistance to mining damages after finishing all post-mining movements being the effect of subsequent extraction, e.g. conducted according to given mining plan of operation.

The issue of the horizontal strain summation in long time intervals is also connected with the problem of the discontinuous linear deformations occurrence. As it is known, the occurrence of these deformations is closely related to the values and direction of extreme horizontal strain values and the course of tectonic disturbances (especially their outcrops on the floor of overburden layers). Such conclusions resulting from detailed analyzes were presented, among others, in the work [13]. It should be noted that methods for forecasting of this type of deformation have not been sufficiently developed so far, which is a serious difficulty in assessing the complex prediction of impact of underground mining works on the surface.

Another important problem regarding forecasts is making calculations of transient values of deformation indices or their extreme values resulting from the movement of the extraction front. This problem is devoted to more attention in the next part of the paper.

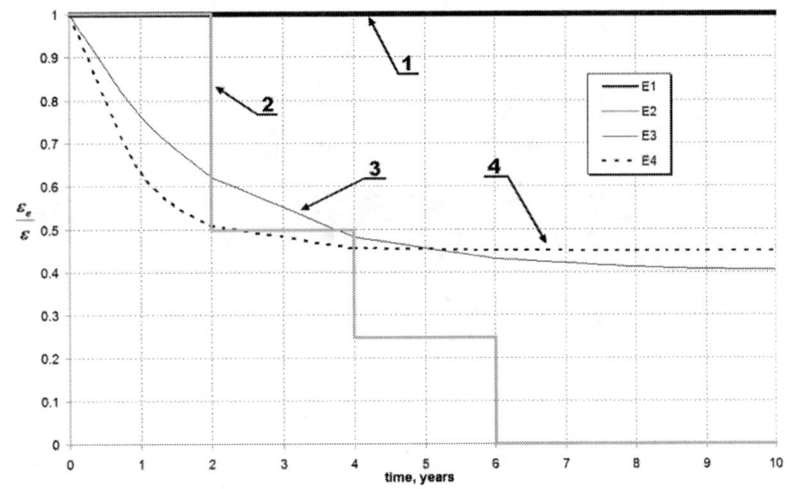

Fig. 1. Time-dependent reduction of horizontal strain influence on the object: 1-by S. Knothe, 2-by B. Skinderowicz, 3-by J. Białek, 4-by J. Kwiatek

2 The exemplary analysis of post-mining deformation distribution over time

Considering the problem concerning the course of deformation state changes over time, it was assumed for calculation purposes, that mining extraction is planned with two longwalls (Fig. 2). Basic information on the mining-geological conditions in the area of considered longwalls is given in Table 1. The table consists of two used data sets, regarding different speed of extraction: 3 m / day and 10 m / day.

All calculations were performed with using the DEFK-Win software [8], based on the theory of S. Knothe [6]. In calculations, the following average (for Upper Silesian Basin mining-geological conditions) values of parameters were used: coefficient of roof control $a = 0.8$, tangent of the angle of main influence range $tg\beta = 2.0$, proportionality coefficient of Avershin's relationship $B = 0.32\ r$, extraction boundary $d = 0$ m.

Analyzing the predicted values of deformation indices, which occur as a result of conducting planned mining extraction at a certain time t, the location of the calculation point in relation to the extraction field should be taken into account. When the object is located at a certain distance outside the horizontal plan of extraction field, the values of deformation indices constantly increase until they reach maxima, which are also the final values. This is illustrated by the example presented below for point No 1, located outside planned extraction fields, near starting roadway, as shown in Fig. 2.

Table 1. Basic data on considered mining extraction

Seam No / extr. rate [m/day]	Longwall No	Extraction start date	Extraction end date	Thickness of deposit [m]	Depth of extraction [m]
207 / v_{03}	718	01-07-2018	20-05-2019	2.5	665
207 / v_{03}	728	21-05-2019	15-09-2019	2.5	670
207 / v_{10}	718	01-07-2018	06-10-2018	2.5	665
207 / v_{10}	728	07-10-2018	11-11-2018	2.5	670

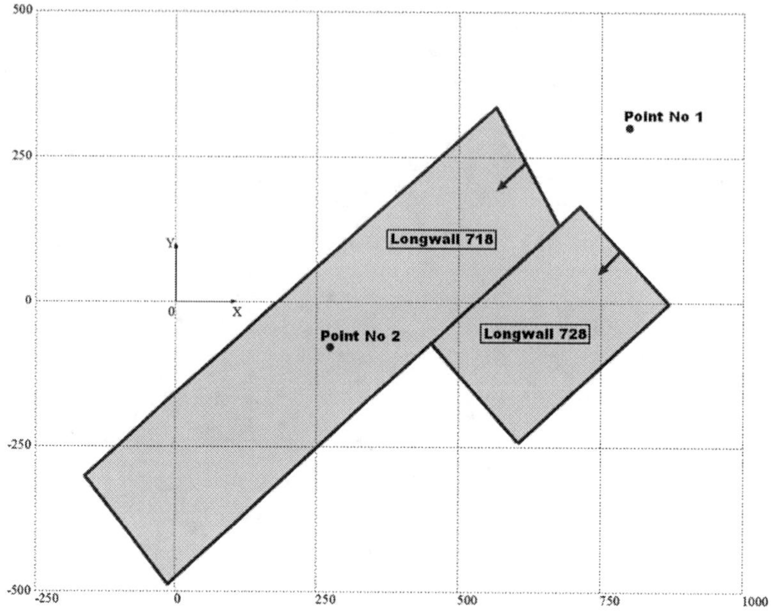

Fig. 2. The sketch of considered extraction fields plan view and location of calculation points

Calculations were performed assuming the immediate influence model - for every considered extraction front position, 100% of influence reaches the surface level instantaneously (without any delay). Fig. 3 shows the course of subsidence over time, Fig. 4 and 5 respectively of maximum tilt T_{max} and principal horizontal strain E_1. As it can be seen from the presented figures, final values of all the above-mentioned deformation indices are also the maximum values for point No 1 located outside horizontal plan of extracted area.

Although subsidence for this point reaches relatively small value of 190 mm, tilt of 2.46 ‰, but horizontal strain of 2.46 ‰ will fall in the range of second category of mining areas (between 1.5‰ and 3‰). Therefore, it can be concluded, that making more detailed predictions of deformation state in case of such location of the object is not obligatory.

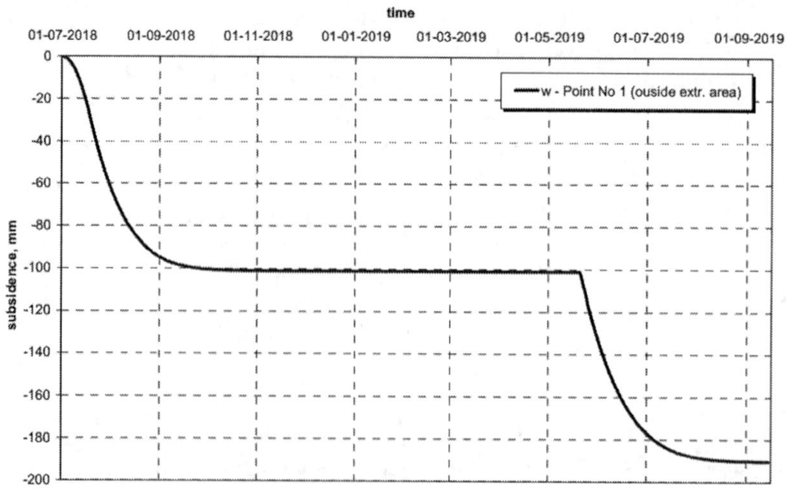

Fig. 3. The course of subsidence over time for point No 1

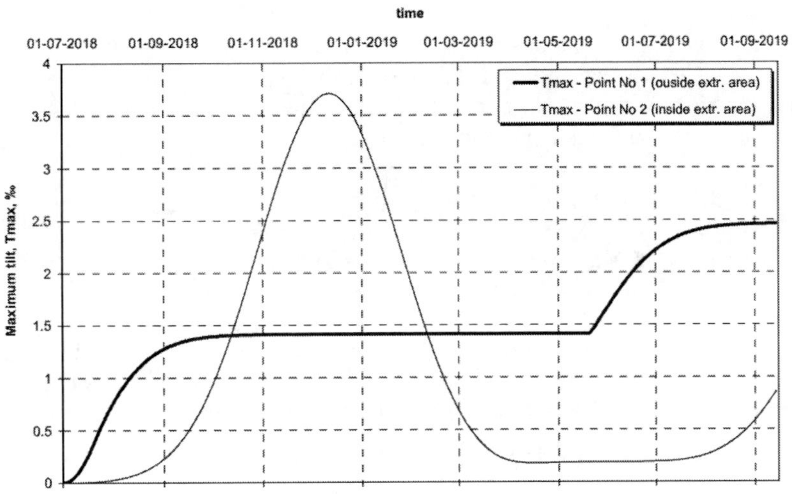

Fig. 4. The course of maximum tilt T_{max} over time for points No 1 & 2

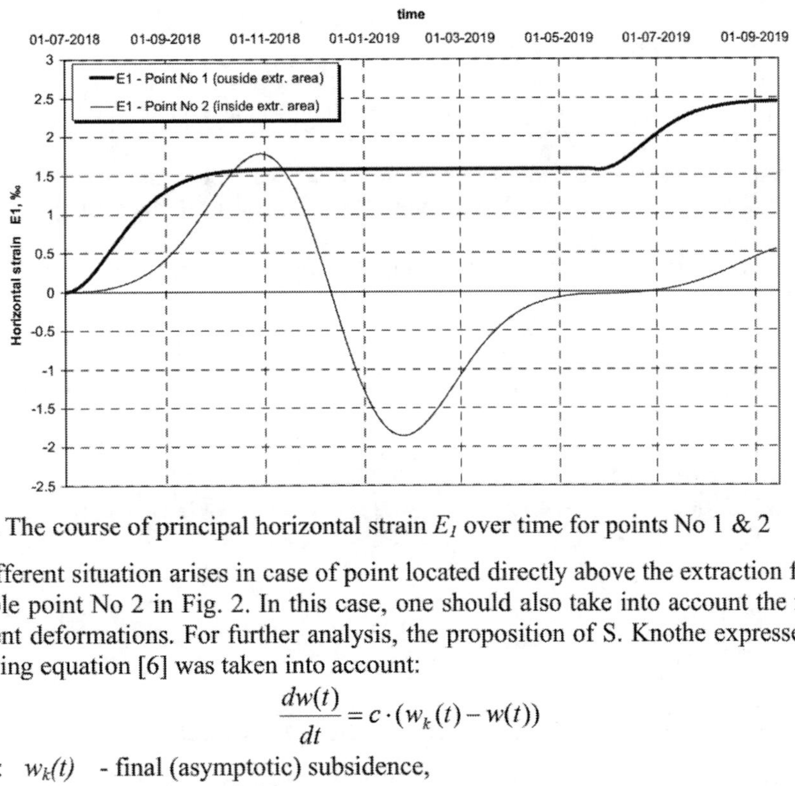

Fig. 5. The course of principal horizontal strain E_1 over time for points No 1 & 2

Different situation arises in case of point located directly above the extraction field - for example point No 2 in Fig. 2. In this case, one should also take into account the model of transient deformations. For further analysis, the proposition of S. Knothe expressed by the following equation [6] was taken into account:

$$\frac{dw(t)}{dt} = c \cdot (w_k(t) - w(t)) \tag{2}$$

where: $w_k(t)$ - final (asymptotic) subsidence,
$w(t)$ - transient subsidence,
c - coefficient of subsidence rate (often called „time factor").

For discretized model (with the assumption: $w_k(t,x) = const. = w_k$) with initial condition: for $t=0 \Rightarrow w(t)=0$, the solution to equation (2) has the form:

$$w(t) = w_k \cdot (1 - e^{-ct}) \tag{3}$$

This solution is wide-spread used in Poland, however other models are utilized in different countries. Detailed description one can find, among others, in [3, 4, 12].

An important problem discussed in [12, 14] is the influence of the extraction advance rate on the course of deformations changes over time. This is reflected in the correlation between the value of coefficient c and the speed of the extraction front progress. Therefore, for further analyzes that take into account the extraction speed, it was decided to use the proposal of an empirical formula presented in [15]:

$$c = 1.463 \cdot \left(\frac{v}{H} \right)^{0,847} \tag{4}$$

where: v - the speed (rate) of extraction front, [m/day]
H - the depth of extraction, [m].

This formula has been worked out on the basis of analyses of survey results taken from Upper Silesia Basin coal mines. The exponent value in the formula (4) is close to 1, but authors concluded that the formula should not be simplified, for the reason that a constant value of the ratio $\frac{c}{r \cdot v}$ (r - the radius of main influence range) would be obtained in such case, so changes in the speed of extraction would not affect the slope of the transient profile of subsidence trough, which stays in contradiction with the results of the measurements.

Calculations of deformation indices values were performed for the following speeds of the extraction and theoretical models:

- Extraction speed v = 3 m/day. Immediate influences ($c \to \infty$) model and delayed influences model described by equation (3) with the value of coefficient c according to equation (4): $c = 0.015$ 1/day.
- Extraction speed v = 10 m/day. Immediate influences model and delayed influences model described by equation (3) with the value of coefficient c according to equation (4): $c = 0.042$ 1/day.

The calculations results of the maximum tilt and maximum horizontal strain are presented together for all of the above variants in figures 6 - 7.

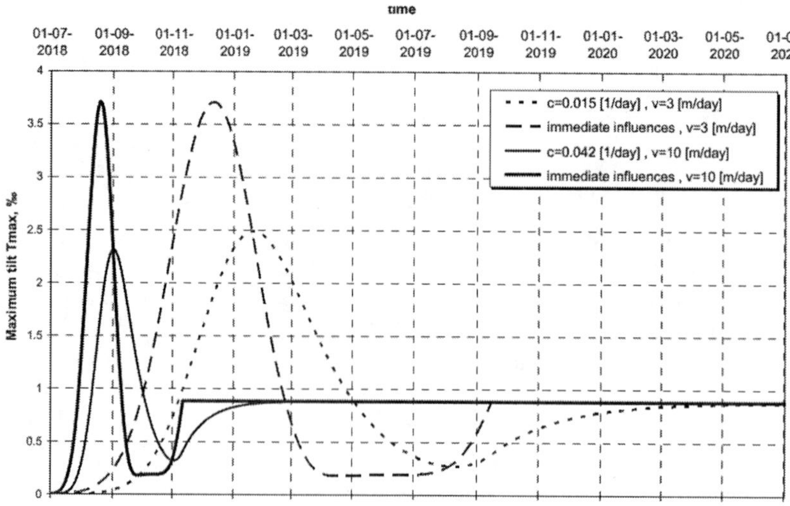

Fig. 6. The course of maximum tilt over time for point No 2

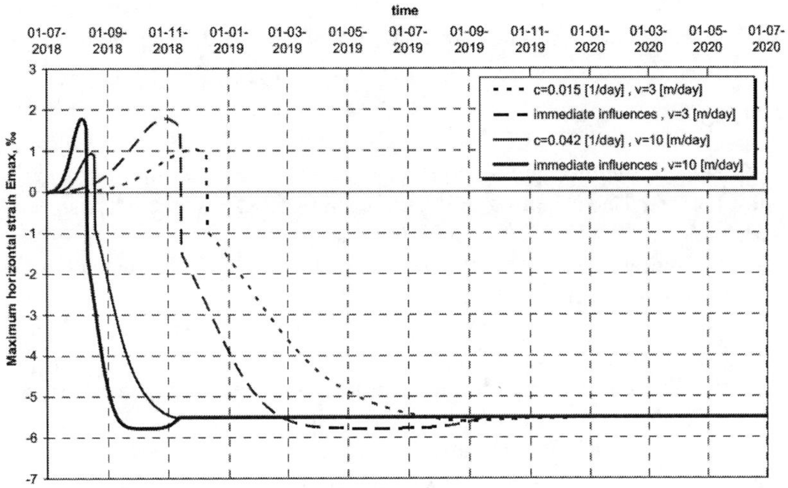

Fig. 7. The course of maximum horizontal strain over time for point No 2

As it can be concluded from the graph showing the course of horizontal strain for the considered calculation variants (Fig. 7), positive values temporarily occurred at the considered object location. This means that the object will subjected to horizontal tensile strain. Ignoring this fact in the analysis of the extraction influence on the object would be a certain inaccuracy, especially considering the low tensile strength of construction elements. Presented results stay in agreement with the analyses published in [12].

A separate problem is the impact of large increase of deformation over time, particularly the subsidence rate. In the case under presented consideration, using the proposal described in [14], there will be subsidence rate from the range of category Ib with the extraction front speed 3 m/day and the category III at the extraction speed of 10 m/day.

3 Conclusions

Analyzes presented in this work allow to make the following final statements and remarks:

1. The problem of post-mining deformations predictions has now become simplified due to the availability of appropriate mathematical models and specialized software. Nevertheless, when making predictions for long time intervals, there are still some doubts, especially considering horizontal strain summation procedure. This problem is also related to the still unresolved issue of forecasting linear discontinuous deformations, that often arise as an effect of large tensile strain. Such deformations, occurring presently more often than in the past, are particularly dangerous for building objects. Thus, the importance of the above problem is still necessary to be emphasized.
2. Forecasts of continuous deformations should include not only determination of asymptotic values of deformation indices, but also their time-extreme values and the rate of their changes over time. Detailed analyzes in the above-mentioned scope should be carried out especially in case of considered object location directly above planned extraction field. Such analyzes allow, among others, determination of transient values of horizontal tensile strain, which is worse tolerated by typical urban construction than compressive one. The quality of the forecast in such cases is decisively influenced by the adoption of the appropriate

values of the parameter characterizing the deformation kinematics - time factor "*c*". In connection with the above, authors recommend using the formula (4).

References

1. M. Chudek: Geomechanics and basics of management of environmental protection in mining and post-mining areas. Publishing House of the Silesian University of Technology. Gliwice (2010) (in Polish).
2. A. H. Goldreich: Die Theorie der Bodensenkungen Kohlengebieten. Verlag von Julius Springer. Berlin (1913) (in German).
3. Kratzsch H.: Mining Subsidence Engineering. Springer-Verlag. Berlin, Heidelberg, New York (1983).
4. P. Strzałkowski: The outline of the protection of mining areas. Publishing House of the Silesian University of Technology. Gliwice (2010) (in Polish).
5. J. Zych, B. Drzęźla, P. Strzałkowski: Forecasting of land surface deformations due to underground mining. Publishing House of the Silesian University of Technology, script No. 1684. Gliwice (1993) (in Polish).
6. S. Knothe: Forecasting the impact of mining exploitation. „Śląsk" Publishing House. Katowice (1984) (in Polish).
7. J. Białek: Algorithms and software for forecasting of mining area deformations. Publishing House of the Silesian University of Technology, monograph series, No. 42. Gliwice (2003) (in Polish).
8. R. Ścigała: Computer aided prediction of rock mass and land surfaces deformations caused by underground mining extraction. Publishing House of the Silesian University of Technology. Gliwice (2008) (in Polish).
9. B. Drzęźla: The description of software for predictions of rock mass deformation due to mining extraction - state of the art. Transactions of Silesian University of Technology, mining series, No. 165. Gliwice (1989) (in Polish).
10. J. Kwiatek: On rheological aspects of the threat of building constructions located in mining areas. Scientific Works of Central Mining Institute, No. 827, Katowice (1997) (in Polish).
11. J. Kwiatek (red.): Protection of building objects in mining areas. Publishing House of Central Mining Institute. Katowice (1997) (in Polish).
12. A. Kowalski: Transient deformations of land surface in the light of prediction quality. Scientific Works of Central Mining Institute, No. 871. Katowice (2007) (in Polish).
13. P. Strzałkowski, R. Ścigała: The causes of mining induced ground steps occurrence – case study from Upper Silesia in Poland. Acta Geodynamica et Geomaterialia. Vol.14 No. 3. Prague (2017).
14. A. Sroka: Dynamics of mining extraction from the point of view of mining damages. Publishing House of the Mineral and Energy Economy Research Institute of the Polish Academy of Sciences, Cracow (1999) (in Polish).
15. P. Strzałkowski, W. Piwowarski, R. Ścigała: The influence of extraction speed on the value of the coefficient of subsidence rate. Submitted for printing (in Polish).
16. A. Kowalski: Problems of prediction of land surface deformation in Upper Silesia Basin. Scientific Works of Central Mining Institute, special issue, V/2007. Katowice (2007) (in Polish).

Influence of continuous deformations and tremors of rock mass on a building. Case study

Piotr Strzałkowski [1,*]

[1]Silesian University of Technology, Faculty Geology and Mining, Gliwice, Poland

Abstract. This work presents an exemplary analysis of the influence of mining exploitations on a building. Continuous deformations of the ground surface in the location of the object were considered. Analysis of the impact of tremors of rock mass on the object was performed. The results of calculations as well as the measurements of surface vibrations accelerations were taken into account. The performed analyses show the influence of a fault on increase of vibrations accelerations.

Keywords: continuous deformations, mining tremors

1 Introduction

Mining is the economy flywheel, which supplies following raw materials: energetic, chemical, metallic and rock. Despite the huge importance of mining, however, there are some negatives of it [1], taking form of its influence of environment, what is mostly associated with so-called mining damage in building facilities [2,3]. This influence shouldn't be exaggerated, because as a result of strict regulations in force in Poland, the necessity of obtaining the consent of the local government administration for the implementation of exploitation plans, damage to buildings is being limited. According to paper [4] they constitute approximately from 1% to 3% of the mining costs. The cost of mining damage removal in Germany in 1996 reached around 350 million DM while the extraction equed 39 million Mg [5]. Among negative forms of influence of mining on buildings there are [2,3]: continuous deformations, discontinuous deformations and tremors of rock mass. The predictions of mining influence on buildings performed routinely should be taking into account all geological factors which can influence the size and deformation distribution. There are many factors, the most significant of them are: exploitations performed previously, hydrogeological conditions and tectonic disturbances. Especially tectonic faults are a common reason of increased values of deformation indices, and even discontinuous deformations. They also influence the increase of values

* Corresponding author: piotr.strzalkowski@pols.pl

of vibrations accelerations of mining ground surface. An example of this kind of influence of the fault zone is presented in this work.

2 Analysis of mining and geological conditions

2.1 Lithology and stratigraphy

The rock mass located in the area of the building consists of layers from following periods:

- Quaternary,
- Neogene,
- Carboniferous.

The overburden consists of quaternary and tertiary layers (Neogene). Carboniferous layers consist mainly of Pleistocene layers, developed as grey and yellow-grey clays stone, sands and fluvioglacial gravels.

Carboniferous layers consist of:

- orzeskie layers – westfal B – group of seams 300
- rudzkie layers – westfal A – group of seams 400
- siodłowe layers – namber B + C – group of seams 500.

2.2 Tectonics

The tectonic fault named Fault V is in the area of the building. It runs from north-east towards south-west direction. The throw slip in the south-eastern direction equals 70 m. The fault outcrop on the carbonic roof runs in the distance of around 150 m in the north-western direction from the building, what is presented in the fig.1.

2.3 Mining exploitations which were performed

The table 1, shown below, presents basic information about exploitations performed since 2005. Values of deformation indices in the location of the building were calculated.

Table 1. Basic information about exploitations performed since 2005

coal seam	wall	beginning of extraction	ending of extraction	thickness of coal seam [m]	angle of slope	depth [m]	distance [km]	direction	a
4132	sc._5	30-09-2005	30-06-2006	1,8	7	690	0,46	W	0,6
4132	sc._4	30-10-2006	30-09-2007	1,8	7	660	0,20	W	0,6
5022	sc._2	01-01-2007	01-01-2008	3,0	10	885	0,49	E	0,8
5022	sc._3	15-07-2016	01-04-2017	3,4	10	920	0,51	SE	0,8
504	sc_1J	02-11-2011	01-03-2012	2,4	8	875	0,49	E	0,8
504	sc_2J	01-04-2014	01-04-2015	2,5	10	910	0,49	E	0,8
506	sc_1	01-04-	01-06-	2,4	5	890	0,53	E	0,8

 https://doi.org/10.1051/e3sconf/20183601009

		2017	2017						

The table gives information about: time of exploitation performed with particular walls, thickness of coal seam or layer, angle (slope) of the seam, depth on which the seam lay in the area of the wall, distance from the building, direction of walls location in relation to the building, value of coefficient of exploitation, the way of roof control. Figure 1 shows the location of performed exploitations in relation to the building, which is marked as number „15".

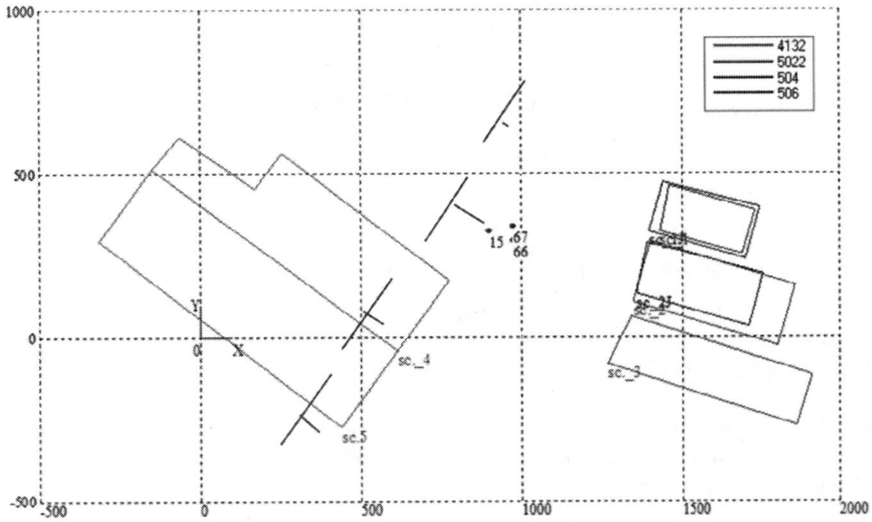

Fig. 1. Scheme of location of selected walls in relation to the object (15) and observing points 66 and 67 in the local coordinate system.

3 Information about the building

The object is a residential building, single-family house with two floors. It was built in 1975 in traditional brick technology. It was protected from exploitation influences by anchoring. The inclination of the building as a result of the impact of exploitations performed in previous years was visible. Many horizontal and diagonal cracks on internal and external walls have been observed since 2015. These cracks occurred mostly where the walls were weakened, so that near doors and windows (fig. 2). No slots in walls have been observed. No linear discontinuous deformations were observed in the area of the building.

Fig. 2. Cracked outer wall

4 Influence of performed exploitations on the building

4.1 Influence of performed exploitations on the building in the light of the results of calculations

Computer programme DEFK-Win [6,7], based on W. Budryk's – S. Knothe's theory [8] was used for calculations of values of deformation indices. The accuracy of the theory has been discussed in paper [9].

The following values of parameters were adopted for calculations on the basis of experience of the coal mine:

- coefficient of roof control – $a = 0,8$ for caving, $a = 0,6$ for caving with locating fly ashes,
- tangent of an angle of main influences – $tg\beta = 2$
- coefficient deciding of values of displacements and horizontal deformations $B = 0,32$

Exploitations performed since 30.09.2005 were taken into account for calculations. Results of calculations of values of deformation indices are presented in table 2, where:

- mining subsidence – w,
- maximal tilts – T_{max},
- maximal horizontal deformation – E_{max},
- maximal vertical curvature – K_{max}.

Table 2. Calculated values of deformation indices

Seam	Wall	w [mm]	T_{max} [mm/m]	E_{max} [mm]	K_{max} [1/km]
4132	sc._5	-0.1	0.00	0.01	0.000
4132	sc._4	-19.3	0.35	0.55	-0.005
502/2	sc._2	-1.5	0.03	0.07	0.000
502/2	sc._3	-0.6	0.01	0.03	0.000
504	sc_1J	-1.4	0.03	0.06	0.000
504	sc_2J	-1.9	0.03	0.07	0.000
506	sc_1	-0.7	0.01	0.03	0.000
Sum		-25.5	0.30	0.66	-0.006

4.2 Impact of performed exploitations on the surface in the area of the building in the light of geodetic survey.

A measuring line, consisting of observing points of numbers from 62 to 67 was set along the street in the area of the building. Observing points number 66 and 67 are located nearest the building, in the distance of around 60 m. Surveys were performed at different intervals since 2007 till 2017.

The course of subsidence of observing points in time is presented in fig. 3, and the increases of subsidence are shown in fig. 4. As the figures show, in time until September 2009 subsidence of observing points, taking values of 86 mm and 85 mm, occurred. Next, the subsidence showed small increases – both negative and positive, what testifies about lack of subsidence. The increases of subsidence of values 21 mm and 27 mm occurred only between September 2011 and January 2012. After January 2012 the increases of subsidence (alternately positive and negative) took the values of measurement accuracy. According to this, it can be assumed that the area of the object was influenced by subsidence caused by mining exploitations latest until January 2012.

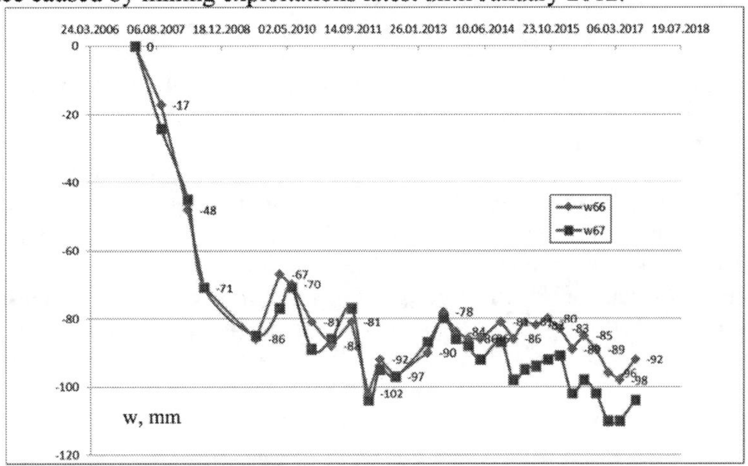

Fig. 3. The course of subsidence of observing points 66 and 67 over time.

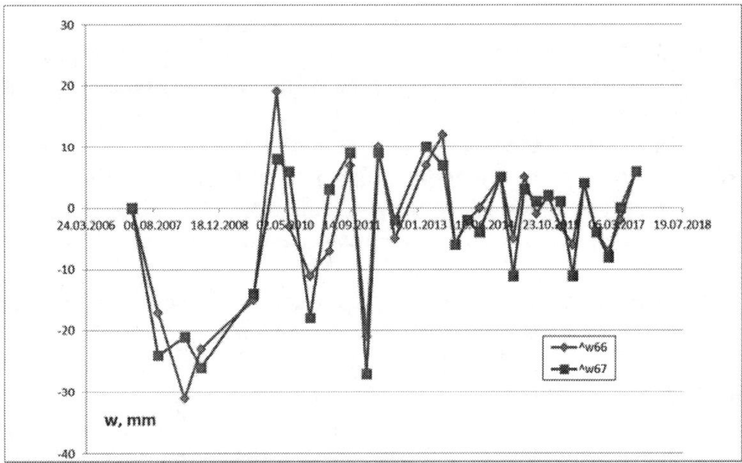

Fig. 4. The course of increase of subsidence of observing points 66 and 67 between surveys.

5 Influence of tremors on the object

The measuring station of the mine records the coordinates of epicentres of tremors and the values of their epicentral energies. Values of vibrations accelerations in the considered place are calculated with usage of the formula of G. Mutke [2]:

$$a = \left[1{,}33 \cdot 10^{-3} \cdot (\log E)^{2{,}66} - 0{,}089\right] \cdot \left[1{,}53 \cdot R^{0{,}155} \cdot \exp(-0{,}065 \cdot R) + 0{,}014\right] \quad (1)$$

where :

a - acceleration of vibrations, mm/s^2

$R = \sqrt{D^2 + 0{,}5^2}$

D - epicentral distance, km

E - energy of the tremor, J

This formula allows calculating the value of accelerations of carboniferous rocks. In order to obtain the surface acceleration, the value obtained from the formula should be multiplied by the so-called amplification factor.
In the light of the results of calculations, no tremor caused accelerations higher (or even equal) 250 mm/s^2 in the location of the building. The mine set a measuring point in the distance of around 180 m from the object. The highest values of accelerations of vibrations registered since 2015 (over 250 mm/s^2) are presented in the table 3. This table also shows values of calculated maximal accelerations and the duration of an intense phase of accelerations of vibrations t_{Ha}. As it can be noticed, the highest registered value was 629 mm/s^2. Values of calculated accelerations for all tremors were lower than measured ones. It is interesting that the tremor which occurred on 02.01.2017 had similar epicentre coordinates and energy to the tremor which occurred on 27.01.2017, and these tremors caused significantly different accelerations of vibrations. The mine rejects the possibility of having made a measurement mistake. These results, so different from expected, were most probably caused by occurrence of the fault zone in the area of the building. The influence of tectonic faults on seismic activity of the rock mass was a subject of many studies [10,11,12,13,14]. A wide overview can be found in the work [15]. The author shows two types of seismic phenomenon. First of them are the occurrences directly related to mining works, which characterise with low epicentral energies. Epicentres of these tremors occur in the area of mining extraction. The second type is related to impact of fault zone by running an exploitation in its neighbourhood. Thus, the combined effect of mining and tectonic factors takes place. Tremors like that characterise with higher energies and their epicentres occur in the area of fault zone. Works like [10] confirm this. Similar cases can be found in foreign literature for example the work [11] shows the case where two tremors have occurred in the gold ore mine in South Africa. The epicentres were located at the depth of around 2 km in the distance of 700m from each other. These tremors have been caused by the mining exploitation and the influence of the fault zone. The influence of tectonics of the deposit on damage of buildings as a result of highly energetic tremor is presented in work [12]. The work presents correlations between accelerations and speeds of vibrations in the location of damaged buildings and distance from fault zone.

It is confirmed by the results of analyses presented in the paper [14], which describes the occurrence of very high amplitudes of accelerations of vibrations in the measuring station in the area of the fault zone. Authors expressed their opinion, that a significant reason for this, was directionality of radiation. It was concluded that tremors characterising with similar energy and distance from measuring station, can cause different amplitudes of vibrations accelerations of the ground surface.

GSI scales are being constantly developed and actualized [16]. Regarding the scale GSI_{GZWKW}-2012-A these tremors can be classified to extents shown in the table. The tremor which caused the highest value of accelerations can be classified to extent II. Due to small distance from the measuring station to the object it can be assumed that the values of vibrations accelerations could have been higher than the calculated ones, also in the location of the building. Therefore there are justified assumptions that the building could have been exposed to maximal accelerations of around 600 mm/s^2, which can be classified to extent II in GSI_{GZWKW}-2012-A scale.

The locations of measuring station and the object have been marked in the figure 5 (small red marks). Also tremors epicentres, data about which can be found in table 4, have been marked. Diameter of the circle represents energy of the tremor. Additionally, for easier distinction, the tremor of energy 4.00E+09 J has been marked with yellow. Epicentres of other tremors were marked with blue colour.

Table 4. Juxtaposition of data concerning tremors, which have caused the highest values of accelerations in the location of measuring station.

Date	E [J]	X [m]	Y [m]	D [m]	PGA survey [mm/s^2]	calculated PGA [mm/s^2]	t_{Ha} [s]	extent GSI
18.04.2015	4.00E+09	20830	-3000	2455	258	230	3,90	I
05.08.2016	4.00E+06	18963	-2122	395	459	151	1,54	I
02.01.2017	8.00E+06	19350	-2790	1096	**629**	142	1,66	II
27.01.2017	8.00E+06	19170	-2900	1070	**300**	143	1,55	I
21.02.2017	2.00E+06	19180	-2830	1017	313	95	1,78	I
14.03.2017	2.00E+06	18460	-2310	331	310	121	1,72	I

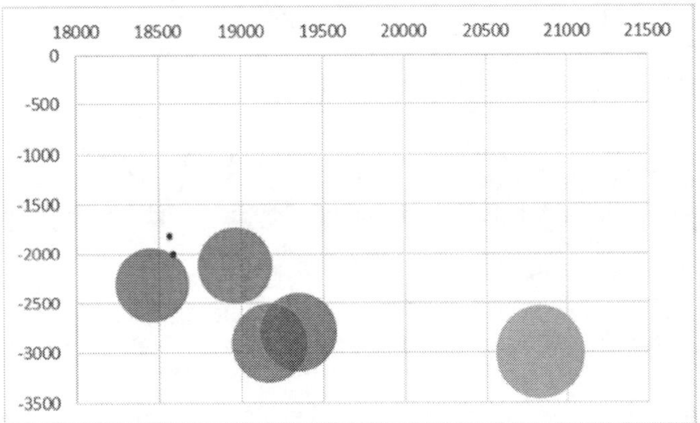

Fig. 5. Location of epicentres of tremors, data about which can be found in table 4 (local coordinate system).

6 Summary and conclusions

Analyses of: mining and geological conditions, results of calculations of values of deformation indices and results of surveys and also impact of the tremors on the surface, performed in this work allow us to conclude:

1. Continuous deformations of the mining area, which developed as a result of exploitations performed in recent years (since 2005) were minor and fit in category I of the mining area. Results of calculations broadly correlate with results of geodetic surveys, which confirm that the surface was not recently influenced by any subsidence. In some cases, however, this compatibility does not occur in tectonically disturbed zone. This is why having access to results of geodetic surveys, which can verify and complete results of calculations, is very important.
2. Analysis of results of calculations and surveys of vibrations accelerations, caused by rock mass tremors, implies as influence of the fault on increase of values of vibrations accelerations of mining ground surface. Formula, which was used, refers to tectonically undisturbed rock mass, which is the reason for observing higher values of vibrations accelerations than calculated ones and even lack of their negative influence on the object. So that also results of surveys of vibrations accelerations proved to be essential for an objective assessment of the impact of mining exploitations on the surface.
3. Mining and geological conditions of the area, including tectonical disturbances, should be considered while assessing the impact of mining works on buildings, since they can influence the size and distribution of deformations, but also they can increase the values of vibrations accelerations on the surface.

References

1. F.G. Bell, T.R. Stacey, D.D. Genske: Mining subsidence and its effect on the environment: some differing examples. Environmental Geology. 40 (1-2) December (2000).
2. J. Kwiatek (red.): Building facilities in mining areas. Publishing House of GIG, Katowice (1997) (in Polish)
3. P. Strzałkowski: Outline of protection of mining areas. Publisher of the Silesian Technical University, Gliwice (2010) (in Polish).
4. O. Kaszowska: Costs of removing damages in buildings located in mining areas. Przegląd Górniczy no. 4 (2005) (in Polish).
5. A. Sroka: Dynamics of mining exploitation from the point of view of mining damage. The Publishing House of the Mineral and Energy Economy Research Institute of the Polish Academy of Sciences, Kraków (1999). (in Polish).
6. R. Ścigała: Computer aided prediction of rock mass and land surfaces deformation caused by underground mining exploitation. Publisher of the Silesian Technical University. Gliwice (2008)
7. R. Ścigała: The identification of parameters of theories used for prognoses of post mining deformations by means of present software. –Archives of Mining Science. vol. 58 iss. 4. (2013)
8. S. Knothe: Forecasting the impact of mining exploitations. 'Śląsk' Publishing House. Katowice (1984) (in Polish)
9. J. Orwat, R. Mielimąka: The comparison of measured deformation indicators of mining area with theoretical values calculated using Knothe's formulas.: International conference of numerical analysis and applied mathematics 2015. ICNAAM 2015, Rhodes, AIP Publishing, (2016).
10. J. Dubiński, K. Stec: Relationship between focal mechanism parameters of mine tremors and local strata tectonics. In: Dynamic rock mass response to mining. The South African Institute of Mining and Metallurgy, Johannesburg (2001).
11. G.F. Hofmann, L.J. Scheepers: Simulating fault slip areas of mining induced seismic tremors using static boundary element numerical modelling. Journal of Mining

Technology. Transactions of the Institutions of Mining and Metallurgy: Section A. Volume 120, Issue 1 (2011)
12. E. Pilecka, R. Szermer – Zaucha : Analysis of the impact of hight – Energy tremor from 29 January 2015 in "Rydułtowy – Anna" coal mine and local tectonics on mining damage in building. Przegląd Górniczy no. 10 (2015) (in Polish).
13. E. Pilecka, K. Stec: Analysis of relation between highly energetic seismicity and gravimetric and magnetic anomalies in the area Upper Silesian Coal Basin. Workshop 2003 from series 'Natural dangers in mining'. (2003) (in Polish).
14. K. Stec, S. Denysenko: Characteristics of tremors of the rock mass observed in the Kłodnicki fault zone– Katowice – Panewniki. Workshop 2003 from series 'Natural dangers in mining'. (2003) (in Polish).
15. K. Stec: Seismic activity of the Upper Silesian Coal Basin – 30 years of the continued seismic observations by Upper Silesian Regional Seismic Network. Workshop 2007 from series 'Natural dangers in mining'. (2007) (in Polish).
16. J. Dubiński, G. Mutke: Mining GSI-GZW Intensity Scale to assess the effects of mining tremors in the Upper Silesian Coal Basin on buildings and people. Scientific Works of GIG, Mining and Environment, Special Edition GIG Quarterly, Katowice (2007) (in Polish)

The influence of para-seismic vibrations, induced by blasting works, on structures: a Case Study

Wacław Andrusikiewicz[1]

[1] AGH University of Science and Technology, Faculty of Mining and Geoengineering, Cracow, Poland

Abstract. Underground mining operations are often associated with the necessity to use explosives. Several hundreds of kilograms of explosives, subdivided into small charges suitable for a specific mining job, are used each time in a blasting operation. In many cases, mining engineers carry out remote central blasting works, which means that all the charges placed at faces are initiated from one control point (usually, a control room in the mine) at the same time. Such coordinated explosions generate para-seismic movements whose consequences can be felt on land surface, with subsequent effects identified in buildings and structures. This paper discusses briefly selected standards applicable to the harmful para-seismic impacts. The author presents the results of the research conducted with the intention to identify harmful effects of the basting works carried out in the "Kłodawa" Salt Mine.

Keywords: underground mining, para-seismic impact

1 Introduction

Each type of mineral extraction operation influences the natural environment in some way. Generally, adverse effects are recorded, with various degrees of intensity. The quality, size, magnitude, and range of impact depends on a number of factors, although the type of mining operations is a dominating factor.

In the case of open-pit mining, we usually observe changes in local landscapes and hydrogeological conditions, forcing subsequent transformations of the local biocenosis. Besides, the method of extraction is also essential here, since blasting works generate para-seismic vibrations that can adversely influence e.g. engineering structures and people.

However, underground mining operations are associated with mining menace, e.g. in the form of land subsidence or sinkholes and rock-mass tremor or collapse etc.

This paper is devoted to the issue of para-seismic vibrations caused by underground rock salt mining. Specific research was prompted by the complaints filed with the "Kłodawa" Salt Mine, which company allegedly contributed to the occurrence of damages in surface structures located around the mining fields where blasting works were carried out.

2 What are para-seismic vibrations

Para-seismic shocks are those induced by human activities. Those include rock-mass vibrations induced by underground mineral extraction (as a result of e.g. chamber ceiling collapse, tremors or rock bursts), or blasting works etc. Moreover, para-seismic vibrations are propagated by heavy vehicle traffic (vehicle impact) or e.g. heavy pile drivers.

Consequently, an underground mine relying on blasting works can be perceived as a potential source generating para-seismic vibrations.

3 Harmful effects of para-seismic vibrations, measured according to scales and standards

Para-seismic impacts cause the vibrations that can be harmful for engineering structures and people. However, how can we assess harmfulness of para-seismic vibrations? That is fairly complicated, owing to the multitude of the sources of vibrations. Different effects are caused by tremor or rock bursts initiated several hundreds of metres underground and still different are those induced by heavy vehicle traffic on land surface, or by blasting works conducted in a quarry. Each of those phenomena will be characterised by specific vibration velocity, acceleration, and frequency. All those parameters should be considered in connection with the source of vibration, as well as the nature of the engineering structures being affected. Apparently, one may think that the issue of the determination of the parameters of the impacts that adversely affect an engineering structure is not complex. However, the number of various classification scales and standards applied to such determinations indicates the opposite.

The Polish Standard [1] is currently applicable in the field under discussion. Actually, it is not the soil or bedrock vibration that is subjected to assessment, but rather the vibrations of the structures being affected by mining operations. The standard in question concerns the dynamic impact scale (SWD) that is presented, for engineering needs, in the form of two nomographs showing the dependence of the vibration velocity on vibration frequency (Fig. 1).

Nomograph SWD-I (left) refers to the buildings with massive structure, small external dimensions in the horizontal plane (not exceeding 15 m each side), single- or two storey buildings, with the height that does not exceed any of the horizontal plane dimensions.

Nomograph SWD-II (right) refers to the buildings not exceeding five storeys above ground, whose height is lower than the double length of the smallest width of the building, as well as low buildings (up to two storeys), all those that do not meet the conditions specified in the SWD-I scale.

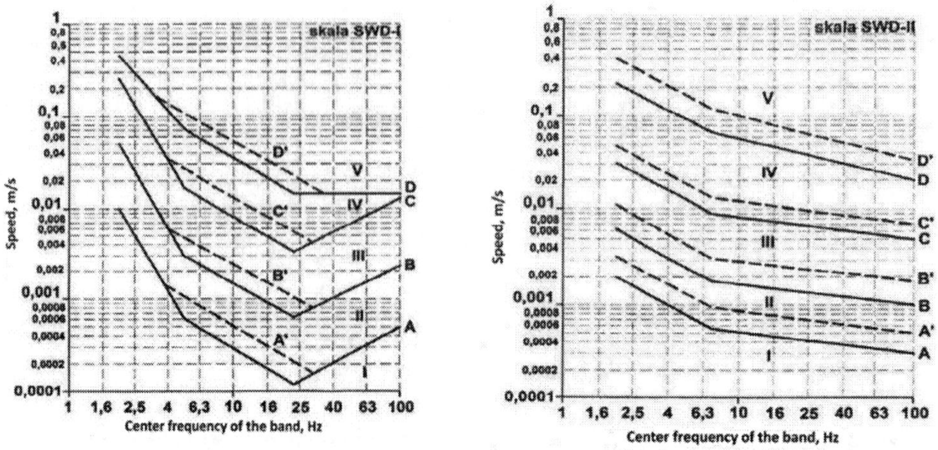

Fig. 1. The lines describing the zones of vibration impact on buildings, in the SWD-I scale (left) and SWD-II scale (right), specified within the following coordinates: central ⅓-octave band frequency – maximum vibration velocity in the band

The nomographs show five harmfulness zones, separated by boundary lines,. The zone boundaries were specified in two options, based on the technical assessment of the building and the types of the substrate (soil, or bedrock) and of vibrations. Assignment to the respective option is effected depending on the dominating number of specific features: Table 1.

Table 1. The features allowing to select the boundary lines of harmfulness zones, within the SWD scales.

Assessment Criteria	Features Allowing to Apply a Boundary	
	Lower Boundary (Continuous Line in Fig. 1)	**Higher Boundary** (Dashed Line in Fig. 1)
Building's Condition	Old buildings, with damages; modified or reinforced buildings	Undamaged buildings, without structural modifications
Building's Material and Structure	Masonry, reinforced-concrete, or stone buildings, poorly structured; no foundations, no ceiling cornices, vaulted ceilings, large or irregular openings in walls, with storey floors at various levels	Solid-brick walls, good workmanship, reinforced-concrete or concrete foundations, massive ceilings binding the walls, with a ceiling cornice
Types of Substrates and Foundations	Low-rigidity substrate (e.g. silty or loose sands), discontinuous or intermediate footing (various heights)	Rigid substrate (e.g. hard silts or clays), flat footing
Types of Vibrations	Long-term or continuous vibrations	Short-term vibrations

The classification of the structures presented in the Polish Standard is limited only to the masonry buildings of up to five storeys. Other types of structures (e.g. reinforced-concrete ones) are omitted, similarly to other engineering structures (e.g. steel towers, reinforced-concrete smoke-stacks etc.).

The vibration issues are regulated in a different way by the German Standard [2]. Synthetic conclusions are presented in Table 2 below.

Table 2. Allowed values of vibration velocity v_i for the assessment of the influence of short-term impact on a building.

Class	Type of Building	Allowed values of vibration velocity v_i [mm/s]			
		Foundation Frequency			Top Storey
		1 ÷ 10 Hz	10 ÷ 50 Hz	50 ÷ 100 Hz$^{*)}$	All frequencies
1	Commercial and industrial buildings and similar structures	20	20 ÷ 40	40 ÷ 50	40 (10)$^{**)}$
2	Residential buildings and their structures and/or similarly used buildings	5	5 ÷ 15	15 ÷ 20	15 (5)$^{**)}$
3	The buildings outside Classes 1 and 2 that are worthy of preservation owing to their resistance to vibrations (e.g. landmarks)	3	3 ÷ 8	8 ÷ 10	8 (2.5)$^{**)}$

$^{*)}$ For the frequencies exceeding 100 Hz, the reference values for 100 Hz may be applied
$^{**)}$ The values in parentheses relate to the impacts that can be considered to be short-term shocks

Besides the standards mentioned here, there are also other classification systems used to determine the threshold of harmfulness, defined as velocity, acceleration, or relocation associated with the intensity of the specific impact, below which no building damages occur, provided that the building's technical condition is assessed as at least good. And again, owing to the multitude and diversity of assessment criteria, such classifications assume various thresholds of harmfulness, distinguishing e.g. between the building's foundation substrates, types and ages of the buildings etc., as well as the sources of vibrations [3].

As we can see, it is not possible to assess clearly the impact of para-seismic vibrations on buildings and structures, especially in the process of measuring the parameters whose values alternate around boundary values. A noticeable increase or drop of such values allows to determine vibration class assessment. Of course, the applicable standard or classification scale will always be a point of reference, deciding whether a given impact is or is not harmful.

4 Case description

The "Kłodawa" Salt Mine is located in the eastern part of the Wielkopolska Region, close to National Road No. 92. The Salt Mine conducts extraction of rock salt in the central section of the Kłodawa salt dome. The mine workings are situated in the deposit, ranging from ca. 425 to 750 m under the ground level. Presently, the works continue on making Level 750 and below available for mining. The workings are cut in seven mining fields. Salt extraction is conducted by the room-and-pillar method. Low chambers are currently cut, in opposition to the previous system of cutting high, cuboidal or cylindrical shapes. Natural hazards occur in the Salt Mine: water encroachment, methane and other gas ejections, and rock bursts. Those hazards are presented in Table 3 [4].

Table 3. List of natural hazards identified in particular mining fields, with the numbers of events (as of 31 December 2015)

		Field No. 1	Field No. 2	Field No. 3	Field No. 4	Field No. 5	Field No. 6	Field No. 7	Outside fields
Water hazard	Degree	III	III	III	III	III	III	III	III
	No. of events	9 (53) [a]	48 (68)	1 (9)	28 (52)	1 (3)	3 (0)	0 (0)	59 (178)
Methane hazard	Category	- [b]	II	- [c]	- [b]	- [c]	- [c]	- [c]	- [d]
	No. of events	91							
Gas ejection, rock bursts	Category	- [e]	III	-	- [e]	-	-	-	- [f]
	No. of events	127, with methane ejection 329, with other gases (e.g. nitrogen, hydrogen sulphide)							

[a] The number of active events (inactive ones in parentheses).
[b] Mining fields belonging to methane hazard Category II, whenever mining includes cutting faces.
[c] Mining fields belonging to methane hazard-free zones, although that determination is cancelled when the methane concentration exceeds 0.1%.
[d] Methane hazard Category II also applies to underground corridors in the areas of the "Chrobry" Shaft, including the Shaft, and the ventilation routes connecting Field No. 2 with the "Chrobry" Shaft.
[e] Mining fields belonging to gas-ejection and rock-burst hazard Category III, whenever mining includes cutting faces.
[f] Category II also applies to the area between the S-W Cross-cut VII at Level 600 m and Field No. 4.

Mining operations are conducted with blasting. Owing to the specific natural hazards, blasting jobs involve ammonite and metanite blasting explosives. About 70 kg of the explosive is used at a face and about 100 kg in a chamber for one blasting job. Nevertheless, there are cases when larger quantities of explosive materials are used. Two blasting jobs are arrange per 24 hours, with several blasting jobs in corridors and chambers each, involving about 1,500 kg explosives in each of the two shifts. Charges are detonated by electronic millisecond-delay fuses.

Various types of measurement devices are used to measure the vibrations of the medium particles, energised by the blasting works. Measurements are conducted, depending on the needs, either in the ground or on the protected buildings or structures. Since direct 3D recording of particle vibrations in the natural medium is not technically possible, the measurement devices break the measured values into three mutually perpendicular components. Two of them being horizontal are marked x and y, where x is a parallel horizontal component and y is the horizontal component perpendicular to the line connecting the place of the blasting job and either the measurement point, or the longer axis of the building in which the vibrations are measured. The third component z is always a vertical axis, being perpendicular to the directions x and y.

The vibration intensity values measured at particular components can be expressed in amplitudes, velocities, or accelerations, as well as 2D or 3D vectors, as the geometric mean of component relocations. The measurement devices allow for data recording, using specialised software for data transfer to a computer for subsequent analysis.

Five measurement stations were established for the needs of this research project. They were equipped with proper vibration gauges designed for the ground environment and the structural elements. Stations No. 1, 3, 4, and 5 were placed on buildings, while Station No. 2 in the ground. Each station was facing the source of vibrations, or the sites of blasting works conducted during two shifts in the salt mine [5].

The list of blasting works conducted at Shift I is presented in Table 4, and those of Shift II in Table 5.

Table 4. List of blasting work locations at Shift I.

No.	Blasting Location	Type of Mining Works
1.	Chamber KSR 8/750*)	rock cutting
2.	Chamber KSR 10/750	ceiling chamber stope cutting
3.	Chamber KSR 14/690	long-hole cutting
4.	Chamber KSR 21/750	rock cutting
5.	Chamber KSR 24/630	surface-attached charge
6.	Chamber Corridor KSR 24/660	rock cutting in the corridor
7.	Chamber KSR 26/660	rock cutting
8.	Chamber KS 1g/660	floor chamber stope cutting
9.	Chamber KS 1g/630	undercutting of a decline ceiling
10.	Chamber Corridor KS 7/720	rock cutting in the corridor
11.	Chamber KSR 11/720	rock cutting
12.	Chamber KS 16A/660	rock cutting
13.	Chamber KS 17/720	rock cutting

*) Working marks: a/b: a – Working Number, b – Mining Level

Table 5. List of blasting work locations at Shift II

No.	Blasting Location	Type of Mining Works
1.	Chamber KSR 8/750	rock cutting
2.	Chamber KSR 10/750	ceiling chamber stope cutting
3.	Chamber KSR 11/720	rock cutting
4.	Chamber KS 19/720	floor chamber stope cutting
5.	Chamber KSR 21/750	rock cutting
6.	Chamber KSR 24/630	½ of a stope + bottom surface
7.	Chamber KSR 26/660	ceiling chamber stope cutting
8.	Chamber KS 3/720	rock cutting
9.	Chamber Corridor KS 7/720	rock cutting in the corridor
10.	Chamber KS 14B/630	ceiling chamber stope cutting
11.	Chamber KS 17/720	rock cutting
12.	Chamber KS 19/720	floor chamber stope cutting
13.	Chamber KS 30/690	rock cutting

The distribution of measurement detectors and blasting work locations are presented in Fig. 2.

Vibration measurements were conducted on the basis of the guidelines specified in the then valid Polish Standard [6], upon placement of the vibration detectors on the building's foundation or load-bearing walls at the ground level (Stations No. 1, 3, 4, and 5) , facing the source of the vibrations. The coordinates of the rectangular grid x, y were assumed as the measurement directions for the horizontal directions and the coordinate z for the vertical direction. Besides, substrate vibration measurements were taken by placing one of the detectors in the ground, with special pins, also assuming the rectangular grid as the measurement direction at Station No. 2. Three-component vibration velocity gauges, type UVS, were use for taking measurements.

 https://doi.org/10.1051/e3sconf/20183602001

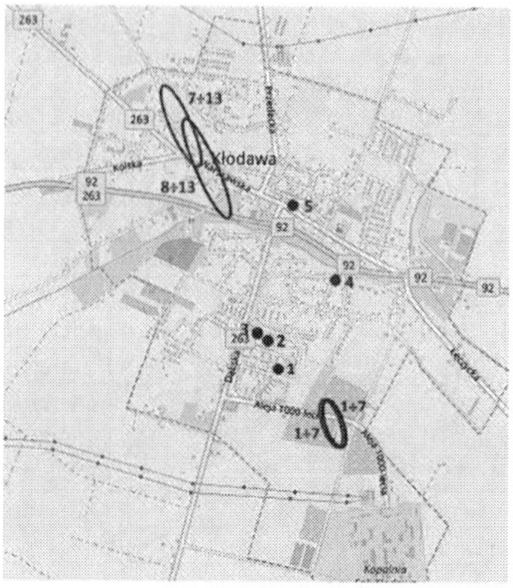

Fig. 2. Distribution of measurement detectors (red) and blasting work locations (blue: Shift I; green: Shift II). Numerals designate the blasting site according to Table 4 and 5

The recorded vibrations could be read only at Stations No. 1 and 5 where equipment was installed to set a low threshold of record initiation (0.10 mm/s), supplemented by abbreviated recordings of only the maximum velocity values, in two-minute periods. Such readings are possible when the vibration values exceed the seismic background noise. In the case of the "Kłodawa" Salt Mine, the low level of vibrations recorded during the measuring sessions allowed for such a reading method. The lists of the results obtained in shown in Table 6 [1].

Table 6. Results of the recorded vibration intensity readings

Blasting Location	**Station No.**	**Record**	**Vibration Velocity, mm/s**			**Frequency, Hz**			**Vector mm/s**
			u_z	u_x	u_y	f_z	f_x	f_y	u_{zxy}
KSR 14/690	1 5	258.741 251.qbar	0.225 0.135	0.100 0.050	0.090 0.060	77.0	22.0	38.0	0.226
KSR 24/630	5	251.qbar	0.060	0.035	0.035				
KSR 26/660	5	251.qbar	0.050	0.035	0.025				
KSR 24/630	5	251.qbar	0.125	0.055	0.030				
KS 1g/660	1	258.qbar	0.089	0.050	0.040				
KS 17/720	5	251.qbar	0.060	0.075	0.081				
KS 30/690	5	251.qbar	0.075	0.035	0.040				

Fig. 3 presents a selected seismogram of the vibrations recorded at Station No. 5 when 10 long holes were blasted in Chamber KSR 14/690, using a 33.5 kg explosive charges in a borehole, detonated with the delay of 25 ms each. The total weight of the explosive used in the series of boreholes amounted to 335 kg, which was about three times more than in the case of standard blasting works conducted on a chamber face. That was the only complete vibration record. The remaining records could be read from abbreviated recordings (*qbar* option).

https://doi.org/10.1051/e3sconf/20183602001

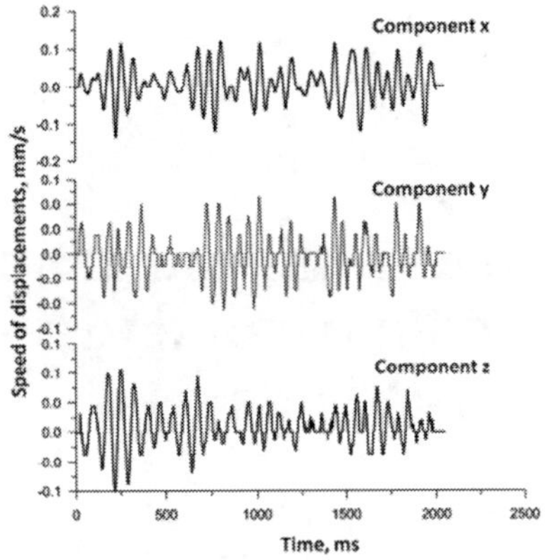

Fig. 3. A building vibration seismogram, Station No. 5

During a measurement session of about 20 minutes (equivalent to the period of blasting works conducted in the mine) the system recorded some vibrations that were not related to the specific blasting jobs. Station No. 5 recorded the vibrations energised by the travel of a heavy truck in the area, see Fig. 4. That event was used to compare the vibration intensity induced by both blasting works and heavy vehicles.

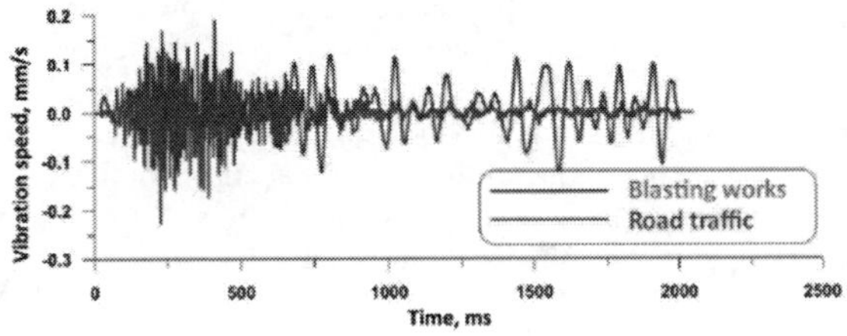

Fig. 4. Vibrations induced by heavy vehicles on the background of the seismic impact induced by blasting works

5 Conclusions

Low intensity of the recorded vibrations makes any analysis practically unproductive. Based on the applicable standards, one can conclude clearly that the para-seismic vibrations induced by blasting and mining works do not have adverse effects on any structures. The levels of the recorded vibrations should be recognized as imperceptible by the structures situated in the vicinity of the salt mine. It is necessary to mention that the recordable vibrations induced by the blasting works can be energised sporadically. However, the traffic, including that of heavy vehicles, is rather continuous in nature when we consider the types of vibration events.

This paper was prepared as part of statutory work No. 11.11.100.005.

References

1. Standard PN-B-02170:2016-12
2. Standard DIN 4150-3:1999-02
3. A. Biessikirski, Przegląd Górniczy **6**, 1-6 (2014)
4. W. Andrusikiewicz, K. Krokos, Przegląd Solny **11**, 57-65 (2015)
5. W. Andrusikiewicz, J. Sieradzki, J. Winzer, Pomiary rekonesansowe drgań parasejsmicznych przy robotach strzałowych w Kopalni Soli „Kłodawa" S.A. (2005)
6. Standard PN-85/B-02170

E3S Web of Conferences **36**, 02002 (2018) https://doi.org/10.1051/e3sconf/20183602002
BIG 2018

The application of Airborne Laser Scaning for identifying old lignite workings – case study: the mine "Borussia" near Ośno Lubuskie (Western Poland)

Agnieszka Gontaszewska-Piekarz[1,*], *Maria* Mrówczyńska[1]

[1] University of Zielona Góra, Faculty of Civil, Environmental Engineering and Architecture, Zielona Góra, Poland

Abstract. The paper presents the possibilities of using data obtained by airborne laser scanning for identifying areas where lignite used to be mined. The technology of airborne laser scanning presented in the paper as and its results have a vast potential in terms of identifying local terrain deformations. The paper also presents the history of lignite mining in the region of Ośno Lubuskie (the north-west of Ziemia Lubuska - western Poland). It describes underground mining in complicated geological conditions (glaciotectonic deformations). The paper is supplemented with historical maps showing the locations of the mines

Keywords: Airborne Laser Scanning, lignite underground mining, terrain deformation

1 Introduction

Lignite was mined in the region of Ośno Lubuskie in the years 1850-1962. There were several dozen mines, most of them operated for a short time and on a small area. In the whole area of Ziemia Lubuska and Lower Silesia there were mostly underground mines. This was due to the geological structure of the deposits (glaciotectonic disturbances) and their irregular location, which made open pit mining unprofitable.

The deposits were mined using the retreat mining method, the workings were not filled, only the shafts were filled, which resulted in different kinds of sinkholes on the ground surface. Those sinkholes appeared mostly during mining activities and shortly after they had been finished. It is possible to assume that in most cases the ground settled directly after the mining activities in a particular mining area had been finished.

The deformations over the old workings can be clearly seen in the open, whereas in woods, in particular in the areas where mining activities were finished in the 19th century, they are more difficult to identify.

* Corresponding author: a.gontaszewska@ib.uz.zgora.pl

In the case of the mines which were closed down before WW2, there is often lack of basic information, e.g. about the mining areas. Therefore, data obtained by airborne scanning can be used. Data obtained in this way enable building high resolution models as well as separating from the cloud of points the information that refers to the terrain cover e.g. buildings and foliage, which makes it possible to build a real terrain model. This is particularly useful for recognizing underground post-lignite mining areas, which at present are located in forest areas [1].

2 Lignite mining near Ośno Lubuskie

The history of mining near Ośno Lubuskie can be traced back on the basis of rather rudimentary archive materials (mining licences, drawings of mining areas from the State Mining Authority), geological documents and mining maps, which are dispersed over different archives.

There are very few documents (geological or historical) with at least any mention of the mines near Ośno Lubuskie. A paper from 1852 includes descriptions of drilling profiles; the author also mentions several mines which operated at that time [2]. A description of lignite mining in Brandenburg, including the area near Ośno Lubuskie, can be found in a book by Cramer [3]. The mines of this region are also mentioned in the most important German handbooks [4,5]. In Polish post WW2 literature there are only brief descriptions (based on German materials) [6,7,8]. The most accurate description of lignite mining near Ośno Lubuskie so far can be found in the works of the author [9,10].

The oldest records of mining areas in the region discussed refer to the mining area "Carl" in Smogóry from 1851 as well as "Bergsegen" and "Rudolph" from 1856. The mine "Carl" was established on 12 October 1855 [3], it gave rise to the later mine "Oskar". This mine, under the name of "Smogóry", operated until the 1960s.

Like in the other regions of western Poland, as a result of the consolidation of small mines and mining areas, in the 20th century there remained only a few operational mines: "Oskar" in Smogóry, "Eduard" in Długoszyn and "Borussia" in Trześniów (Fig. 1.).

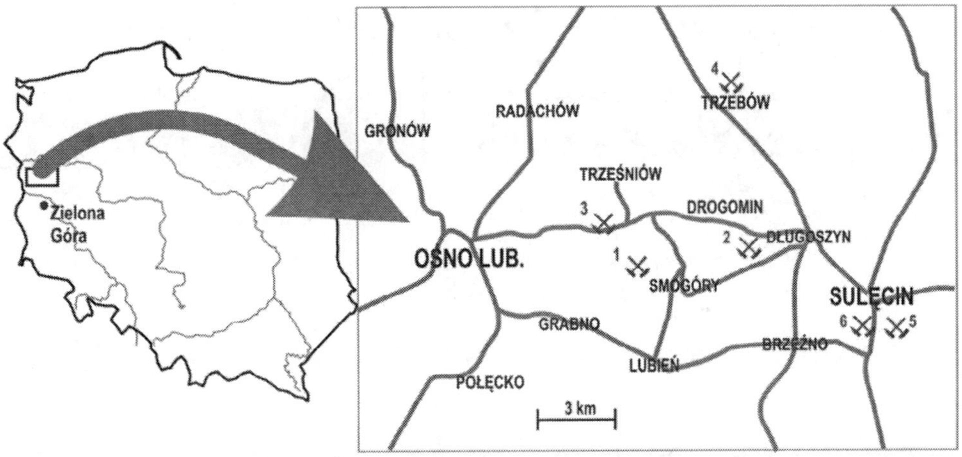

Fig. 1. A map of the most important mines near Ośno Lubuskie. 1 – Oskar/Smogóry, 2 - Eduard/ Długoszyn, 3 – Borussia, 4 – Gute Hoffnung, 5 – Phönix, 6 – Alexander

The lignite deposits mined near Ośno Lubuskie are located in glaciotectonically transformed structures. These processes were particularly intensive in marginal ice zones. The glaciotectonic structures of western Poland appeared mainly during the second

https://doi.org/10.1051/e3sconf/20183602002

glaciation of the San river [11] or the second glaciation of the San river and the glaciation of the Oder river [12]. The ice sheet of the subsequent, younger glaciations could have caused further structural changes.

Before WW2 mining activities were conducted only in gaciotectonically disordered structures, where lignite deposits, initially located at great depths, had been "pressed up" towards the surface, which made them accessible for mining.

The deposit mined near Ośno was the "I - środkowopolska grupa pokładów" or "Oberflöz" in Pre-WW2 German literature (now 1. Lausitzer Flöz). This deposit consisted of over a dozen parallel, longitudinal anticlines, which were called saddles. The saddles were partially eroded by the subsequent glaciers and are mostly located under a thin cover of tills. Mining activities were conducted in the places where the deposits were at the shallowest depth, most often in saddle joints (Fig.2). The lignite deposit originally located at about 0 m above sea level near Ośno Lubuskie was upthrust to about 100m above sea level [13]. The deposit was accessible by shafts or descending galleries as well as horizontal galleries, which were also later used for transporting the lignite. The shafts were usually built at the highest points of the saddles. The room and pillar retreat mining method was used. This method consists in removing lignite from rooms and leaving safety pillars of unremoved lignite [4,14]. The lignite was transported (usually manually) into the tunnel. Additional ventilation shafts were also built at the end of the tunnels, sometimes also additional transport (mining) shafts.

In the case of steep lignite deposits, which were very common, the mining tunnels were built one above the other, moving downwards. Those tunnels were connected by chutes, which were used to dump lignite to the lower tunnels and then to the main tunnel through which the output was transported to the shaft. The height to which lignite was mined in a room did not usually exceed 5 m, and the dimensions of the room were from 3x4 to 4x5 m. While the lignite was removed, the ceiling was secured with a wooden support. That wood was stripped after the mining work was completed, which caused the overlayer to collapse. After the ceiling had collapsed, work on another room started [13,14].

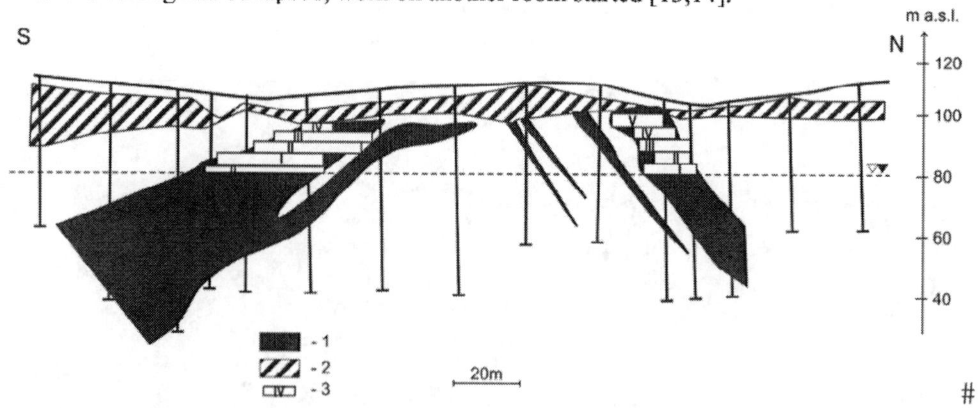

Fig. 2. A simplified geological cross-section of one of the folds of the "Smogóry" mine. 1 – lignite bed, 2 – till, 3 – mining level [10]

3 The mine "Borussia" – preserved historical records

The registration of the mining area "Oscars Segen" of 29 November 1855 is regarded as the beginning of the mine "Borussia". In 1874 the mine "Oscars Segen" was consolidated with other mines and mining areas and given the name of "Borussia" [13]. In the 1920s the mine "Borussia" was included in the neighbouring mine "Oskar".

Lignite was mined to the south and north of the railway line Ośno Lubuskie – Sulęcin (Fig.1., Fig.3.). The shafts marked as I and II can be seen on topographic maps (from 1899 and 1927), and shafts marked as H, K and J on a geological map from 1905. No mining map with these shafts has been preserved. Shafts H, K and J were 30-40 m deep. In each of these shafts two beds were mined. Their thickness was about 3.0 – 3.5 m and they were separated from each other with a layer of sand with a thickness of about 8-10 m. There was one more bed below, which was mined only through shaft H [5]. It is quite possible that on the topographic maps the names were changed: shaft H to shaft II and shaft J to shaft I (Fig.3., Fig.4.).

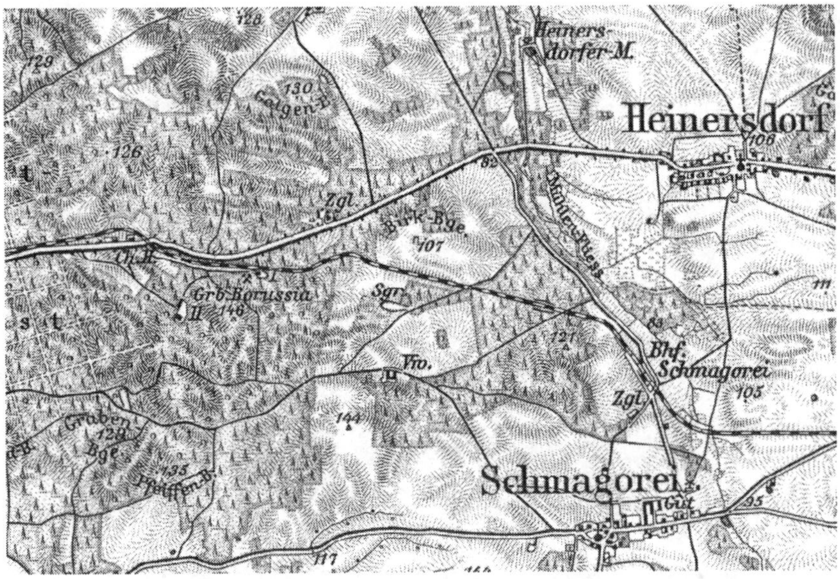

Fig.3. The map from 1899. Shaft I and II of "Borussia" mine are marked

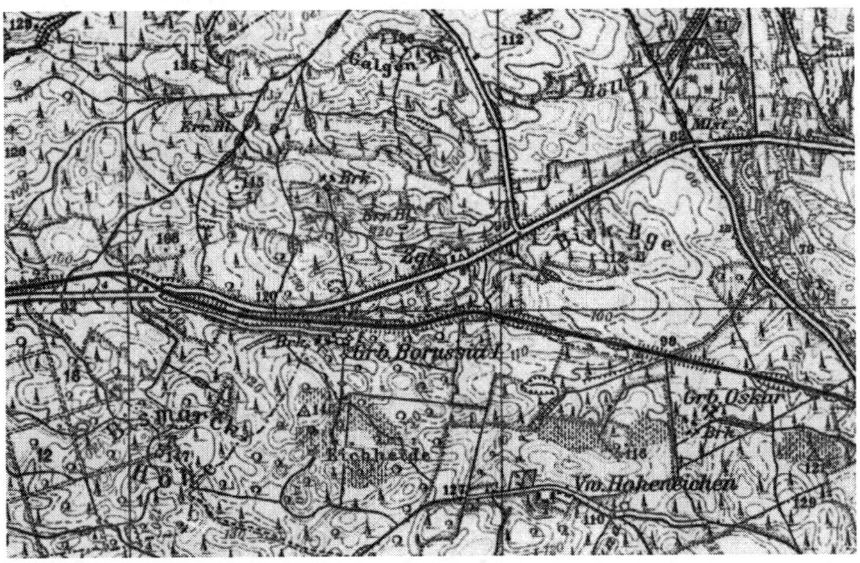

Fig. 4. The map from 1927. "Borussia" shafts: I and Otto are marked

A German pre-WW2 geological map also mentions the location of saddles (folds) in the mining area of "Borussia" and shows other shafts. It is not known whether those shafts belonged to "Borussia" or to the other, smaller mines.

Detailed information has only been preserved about the lignite deposit mined through the shaft "Otto", which operated in the years 1924-1932. It was 46.39 meters deep and was located about 500 m north of the railway line. Two beds were mined in it: the lower one (II Flötz), whose main tunnel was located about 66-67 m above sea level and the upper one (I Flötz) with the main tunnel located about 91 m above sea level. Complete 1:1000 mining maps have been preserved for both beds. Apart from the shaft "Otto", the deposit was accessible by a descending gallery located in the eastern part of the bed.

The output from the shaft "Otto" was transported by rail to the railway siding where it was sifted and briquetted. The briquette factory was located directly south of the siding. The mined lignite deposit crossed the Ośno –Sulęcin road for which a safety pillar of about 40 m was built. Only transport tunnels were built near the safety pillar.

4 Airborne laser scanning

The ALS (Airborne laser scanning) is one of the remote sensing technologies [15,16].

The ALS (Airborne laser scanning) system consists of two segments: the on-board segment and the on-ground segment. The on-board system consists of a laser rangefinder, a plane positioning system based on the GNSS (Global Navigation Satellite System), the INS (Internal Navigation System), a digital camera, a data recorder and the FMS (Flight Management System). The on-ground segment includes: a land based GNSS reference station and specialist software, which makes it possible to process data obtained as a cloud of points. This technology is often called a laser radar or LiDAR (Light Detection And Ranging).

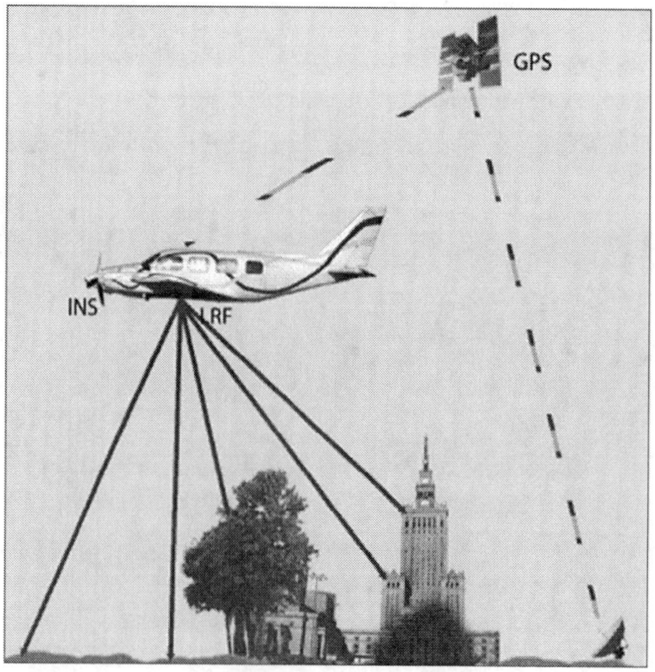

Fig. 5. The technique of airborne laser scanning, prepared by the author, based on [1]

The technique of data acquisition by airborne laser scanning is based on laser measurements of the distance between points located on the terrain surface and a flying plane. For this purpose a laser rangefinder is used. The rangefinder is also connected to the GNSS, which makes it possible to find the position of the flying plane. Moreover, on the basis of the points obtained with the INS, the angle of inclination and the velocity of the platform with the scanner head are determined [1,17]. All this information makes it possible to determine the coordinates (X,Y,H) of points located on the terrain surface (Fig. 5). A cloud of points obtained in this way, being a quasi-continuous metric terrain model including objects located over and on the terrain Surface, is an intermediate product (a starting point) for further analyses. In order to use the data obtained as a cloud of points, it is necessary to process them, and the basic method of processing a cloud is classification, which consists in assigning an attribute to each point. In this way points representing different categories can be separated from the whole, e.g. points located on the terrain surface, buildings, low, medium-height and high plants as well as other categories defined according to the user's needs. This approach makes it possible to prepare final products, which could be: digital elevation models: a digital land cover model (DLCM), a digital terrain model (DTM), spatial models of buildings, cities or vector maps e.g. of communication routes, coastal zones, watercourses or power lines.

5 Research area

The mining areas of the mine "Borussia" are known from the preserved topographic maps (mainly the location of the shafts), the mining maps, (only the deposit mined through the shaft "Otto") and the geological map. The geological map from 1905 shows the location of shafts (probably ones belonging to different mines) and the strike of the anticlines, which is consistent with the mined deposits (Fig. 6). As a result of the use of the room and pillar retreat mining method, very characteristic funnel-shaped sinkholes appear on the terrain surface. Near the villages of Smogóry and Trześniów rows of such sinkholes can be clearly seen. They precisely mark the mining area. The picture is more accurate in the case of the newer workings.

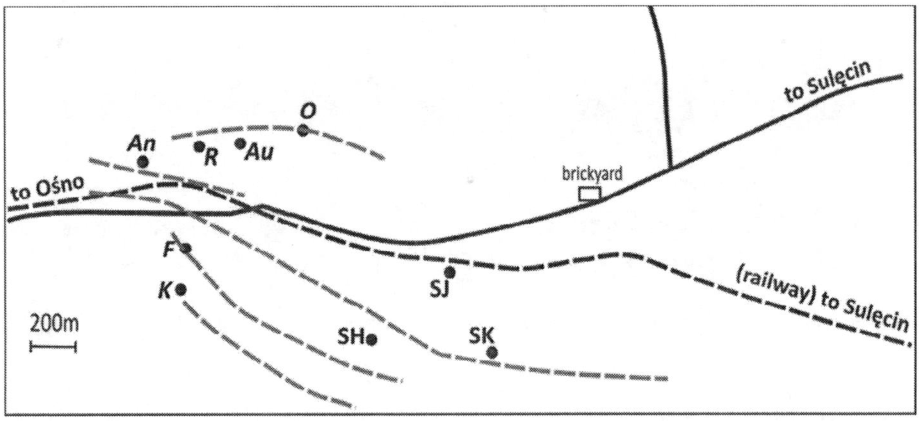

Fig. 6. A sketch of the "Borussia" mine surroundings, the end of the 19th century (based on the geological map). Mining areas: "Borussia" shafts: SH – shaft H, SK – shaft K, SJ – shaft J, other shafts (mines?): An – Andreas, R – Rudolf, Au – August, O – Otto, F – Franz, K – Konrad. The strikes of lignite anticlines have been marked [10]

The data obtained by airborne laser scanning made it possible to identify a number of details, including no longer existing structures. In the case of the shaft "Otto", it is possible

https://doi.org/10.1051/e3sconf/20183602002

to clearly see the former railway line connecting the shaft with the railway station and then with the briquette factory, as well as the location of the briquette factory itself (already demolished in the 1930s). It is also possible to see the descending gallery, which was located in the eastern part of the deposit (Fig.7,8).

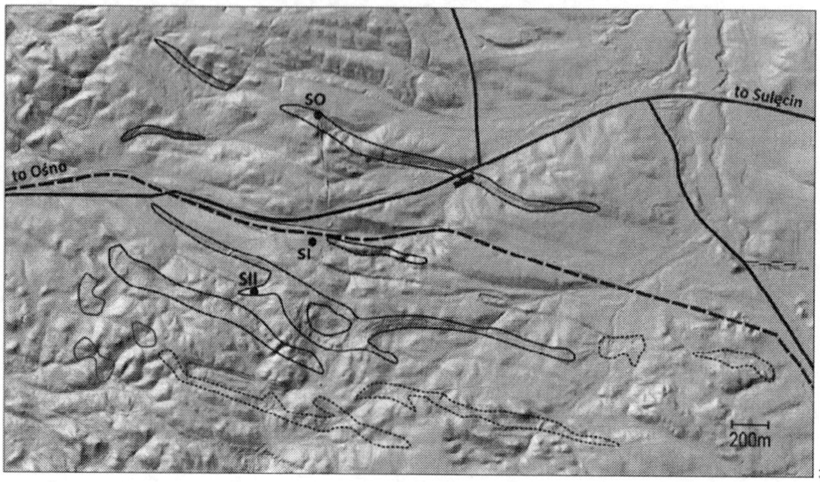

#

Fig. 7. A numerical terrain model of the mine "Borussia". Explanations: SO – shaft "Otto", SI – shaft I, SII- shaft II. The deformation areas corresponding to the mining area have been marked. The broken line marks the deposits mined by the mine "Oskar", the red line marks a probable mining area (19th century).

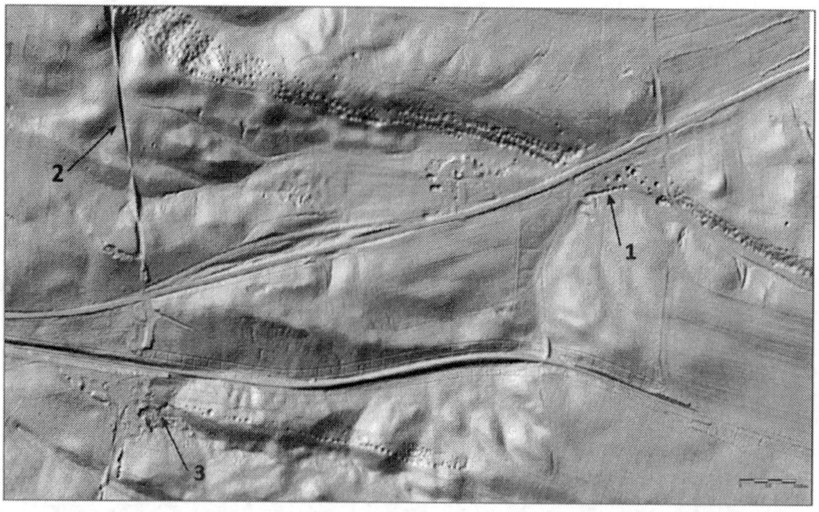

Fig. 8. A numerical terrain model of a part of the mine "Borussia" - Otto shaft. 1- descending gallery, 2 – railway between shaft and briquette factory and railway station, 3 - briquette factory

6 Conclusion

The above figures clearly show that airborne laser scanning can be very useful for identifying terrain deformations. This is particularly important in the case of places where direct measurements are difficult, such as forests. The room and pillar retreat mining method, which was used in the underground lignite mines in western Poland, caused very typical terrain deformations – a row of craters. They make it possible to precisely determine the mining area, both with traditional geodesic methods and by airborne laser scanning. The depressions formed in the 20th century in forests areas have been preserved in excellent condition, and the older deformations (from the 19th century) not so well, but it is still possible to identify them. Airborne laser scanning is a technology with a vast potential for identifying post-mining areas, in particular where historical data are missing. Data from airborne laser scanning require analytical instruments that enable data visualization as well as algorithms for processing clouds of points.

Data from airborne laser scanning make it possible to precisely represent terrain deformations in the form of a numerical terrain model, which undoubtedly is an advantage in the case of forest areas.

References

1. Kurczyński Z., Fotogrametria, Państwowe Wydawnictwo Naukowe, Warszawa (2014)
2. Plettner F., Die Braunkohle In der Mark Brandenburg. Ihre Verbreitung und Lagerung. Berlin (1852)
3. Cramer H., Beiträge zur Geschichte des Bergbaues in der Provinz Brandenburg, Heft 1, Kreis Sternberg, Halle, (1872)
4. Klein G., Handbuch für den deutschen Braunkohlenbergbau, I Band. Verlag von Wilhelm Knapp, Halle (1927)
5. Pietzsch K., Die Braunkohlen Deutschlands. Berlin (1925)
6. Poborski C., Geologia złóż węgla brunatnego. [in:]: Węgiel brunatny w Zachodniej Polsce, Katowice (1949)
7. Suszyński K., Przegląd Górniczy, 1-4 , 33-41 (1946)
8. Żaba J., Prace Naukowe Uniwersytetu Śląskiego 169, 108-142 (1977)
9. Gontaszewska A., Hereditas Minariorum, 2, 51-65 (2015)
10. Gontaszewska A., Zeszyty Naukowe Uniwersytetu Zielonogórskiego, 42, 89-101 (2016)
11. Mojski J.E., Ziemie polskie w czwartorzędzie. Państwowy Instytut Geologiczny, Warszawa (2005)
12. Urbański K., Zeszyty Naukowe Uniwersytetu Zielonogórskiego, 129, 159- 173 (2002)
13. Gontaszewska A., Przegląd Górniczy, 10, 1-8 (2015)
14. Grumprecht A., Grundlagen der Braunkohlenbergbau. Verlag Wilhelm Knapp, Halle (1948)
15. Pilecka E., Piątkowska A., Graniczny M., An application of the remote sensing technology to the high-energy seismic activity assessment, on the example of the Upper Silesian Coal Basin (USCB), Polish Geological Institute Special PapersVolume 20, 92-98 (2005)
16. Pilecka E., Szwarkowski D., An application of the ground laser scanning to recognise terrain surface deformation over a shallowly located underground excavation, E3S Web of Conferences, Vol. 24, 1-8, (2017)
17. Wężyk P., Podręcznik dla uczestników szkoleń z wykorzystaniem produktów LiDAR Główny Urząd Geodezji i Kartografii, Warszawa (2014)
18. Jaros J., Słownik historyczny kopalń węgla na ziemiach polskich, Katowice (1984)

The use of two-axis high precision inclination sensors in determining headframe deflection

Wojciech Jaśkowski*[1], *Mateusz* Jabłoński[1], *Małgorzata* Ulmaniec[2], and *Krzysztof* Paraszczuk[3]

[1]AGH University of Science and Technology, Krakow, Poland

[2]Cracow University of Technology, Faculty of Civil Engineering, Krakow, Poland

[3]District Mining Office in Krakow, Krakow, Poland

Abstract. Reliability of transport equipment in the shaft depends, among other things, on the accuracy of the vertical foundation of the headframe together with elements of the hoisting equipment over the shaft. Any deviations beyond the installation and movement tolerances may cause incorrect or even dangerous operation of the hoisting equipment in the shaft. Therefore the headframe is subjected to periodic inventory measurements, which prevent the movement in the shaft and the smooth operation of the whole underground mine. The Wieliczka Salt Mine developed a project for the installation of precision Nivel 220 two-axis inclinometers on Kinga and Daniłowicz headframes. The paper summarizes the initial conclusions resulting from the first year of the system operation and indicates the directions of its development.

Keywords: precise, inclinometers, headframe, deflection

1 Inventory measurements of headframe in Polish law

Inventory measurements of headframes and their range depend on the type of deformations occurring in the facility and should meet the requirements set for them in the relevant regulations. Diversified construction of these objects, as well as the purpose of inventory measurement, generally require the development of individual programs for these measurements. In particular, when there are external causes that can cause headframe deformations, for example effects of mining exploitation carried out in a shaft pillar – then the inventory measurement program should include a broadly understood control of geometry changes of particular construction axes defined in Polish Standards [1]. In such cases, unevenness of depressions at the foundation points of the headframe may lead to warping of the structure, and thus pose a threat to the use it.

* Corresponding author: jaskow@agh.edu.pl

On 9 June 2017, pursuant to the Regulation of the Minister of Energy of November 23, 2016, on the detailed requirements for the operation of underground mining facilities [2], the Regulation of the Minister of Economy of 28 June 2002 on health and safety, traffic management and specialized fire protection in underground mines [3] has been repealed. The previous Regulation left the measurer a lot of doubts, especially regarding the deflection of the headframe, mainly due to the provisions in §469.1 [3]. The problem was that the measurement should be subject to the stem of the headframe treated as an extension of the guides from the shaft, meanwhile the measurers are required to extrapolate its declination to the height of the rope pulleys, which are located on the head, the element independent of the stem. This means that the stem of the headframe has its axis, independent of the axis of the head (available on figure 3) [4, 5]. In 2017, the legislator attempted to clarify the above provision (§542.1.2 [2]), introducing another concept of "axis of the headframe", also unknown in the Polish Standard [5], so the change of the regulation did not change anything. At the same time in §542.3 there was a record enabling further use of the headframe despite exceeding the above deflection with the consent of the Chief of the Mining Facility Movement, based on the opinion of the expert.

2 Measurements of headframe's verticality changes

Measurements of the verticality of the headframe are sometimes very laborious due to the various constructions of headframes and pose a lot of technological difficulties. Often, it is sufficient to measure relative changes in headframe's verticality to assess its technical condition or the influence of external factors affecting it. The difference in the measurements of relative changes in the verticality of the headframe in relation to the absolute measurements lies in the lack of the need to determine the mutual position of the measurement points in relation to the vertical axis of the headframe. Omitting the designation of these elements greatly simplifies these measurements and allows the use of all methods regardless of the construction of the tower. There is also no need to stabilize the control points exactly in the plane of the headframe main axis.

Application of the above-mentioned principle leads to the fact that the designated coordinates of control points, located on the headframe, can be used to determine the state of its position in the vertical, after determining the initial eccentrity. This means that the cyclic positioning of these characters can be interpreted only as a periodic change in the position of the structure in the vertical (equation 1, fig. 1).

$$P_i = M_i + dP_i \tag{1}$$

where: P_i – total resultant deflection vector of main axis on level „i"; M_i – initial resultant deflection vector (initial eccentrity) of main axis on level „i"; dP_i – periodic changes of main axis on level „i".

To obtain the angular value of the tower inclination (η_i), the vector should be divided by the height (H_i) of a particular level, relative to the basic girder of the shaft. Due to small values of vectors on particular levels in relation to their height, the function of small angles could be used, determining the values of deflections from the equation 2 for changes in total deflections of individual levels and 3 for periodic changes of deflection of individual levels.

$$\eta_i = \frac{P_i}{H_i} \tag{2}$$

$$d\eta_i = \frac{dP_i}{H_i} \tag{3}$$

where: H_i – height of level „i", according to regulation [2]

It follows from the above that observations that determine the coordinates of control points in a cyclical manner allow only to determine periodic deviations of the tower structure from the vertical. Determination of initial inclination M_i is based most often on inventory measurements of characteristic axial points of the structure. This makes it possible to determine the total deflection of the construction axis from the vertical. It should be remembered that the headframe's inclination value determined in this way - in the form of a displacement vector at a given height are – only information related to the moment of the measurement. Therefore, such measurements should be carried out in the absence of any additional factors affecting the structure.

When measuring verticality or changes in verticality of the headframe, it is important to determine the changes of the headframe during its operation. Also, the determination of the impact of external conditions, such as solar radiation, wind pressure and impact of mining shocks, allows for a more complete picture of the behaviour of the structure. Determination of the maximum values of dynamic changes in the position of the headframe, caused by the above-mentioned factors, allows taking into account these values in the measurements of the structure verticality. Thus obtaining the actual values of the tower deflection, can then be compared with the limit values.

The solution to these problems is the use of measuring methods that provide quasi-continuous measurement of changes in the hadframe's deflection. These methods, included in the telemetric methods, automate the measurement process, ensuring the receipt of information sets characterizing at each programmed moment. The measurement results are used further to calculate and evaluate the process. Most of telemetry systems used in geodesy are based on electrical systems [6]. The signal source in such measuring system is the sensor, on which the measured quantity acts directly. In the case at hand, this is the change in inclination. Inclinometers are tools for measuring instantaneous and continuous changes in structure inclination to determine the deformations that occur on them. The use of precise inclinometers allows to determine from obtained changes the actual values of the inclination of headframe's axe. The arrangement of inclinometers on the construction elements of the headframe, located at different heights also allows defining the deformation of the headframe structure [7]. The use of two-axis inclinometers (eg Leica's Nivel 220 [8]) allows simultaneous determination of these values in two perpendicular directions [9-11].

The data recorded by sensors show the current slope of the plane on which they are located. Recalculation of data from inclinometers on the values of declination changes in headframes is made according to the equations 1 and 4 (fig.1).

$$dP_i = D_i \tan\alpha \tag{4}$$

where: dP_i – periodic changes of main axis on level „i"; α – the inclination angle obtained from the inclinometer; D_i – height of inclination sensor location.

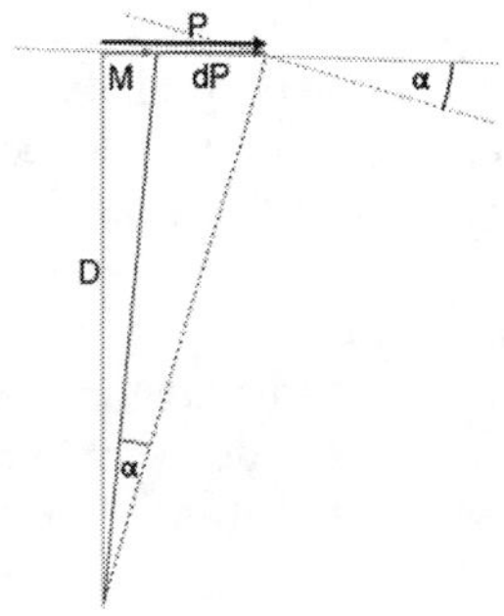

Fig. 1. An algorithm for converting data obtained from inclinators into values of changes in the headframe's inclination, variables according to equations 1 and 4.

3 Measurement system for continuous monitoring of Daniłowicz and St Kinga headframes in the Wieliczka Salt Mine

To support the inclinometers installed on the headframes that ensure full use of these devices in continuous monitoring of changes in headframe's inclination, a device design has been developed (Fig. 2). The designed device has been equipped with a microcontroller, whose task is to acquire data from two-axis inclination sensor Nivel 220 (Leica Geosystems AG) through the RS485 interface. The data is archived on a removable SD memory card installed in the device. The microcontroller was also connected to the WiFi transmission module, which enables the transfer of data from the inclinometer via a network with a computer installed in the mine's measurement department, but due to technical reasons this form of communication was abandoned. The device is powered from the 230V mains supply to the level of pulleys in the TNS network configuration.

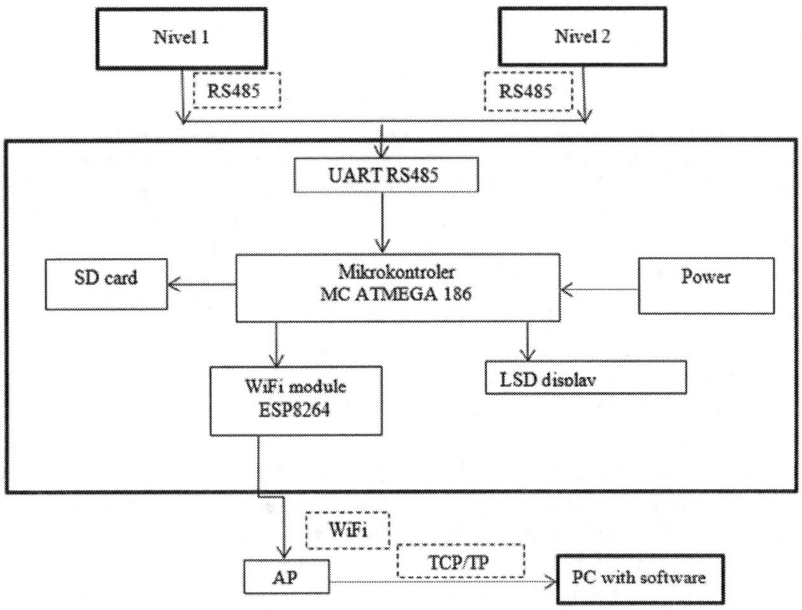

Fig. 2. Diagram of the device for recording and transmitting measurement data.

Obtained data are transferred to the recording device via the RS485 port. On the computer (and at the same time data server) specially created software has been installed. That allows converting data from inclinometers into headframe inclination changes and their visualizations. The recording of data takes place in the recording device on the SD card with a 68-second interval.

The Daniłowicz and St. Kinga headframes have a similar height of the rope pulleys, 22.3 m and 23.5 m, respectively. Their maximum allowable deflection is about 45mm. Regarding both towers it was decided about the need to look independently at the stem and head (fig. 3). One inclinometer was installed on the rope pulley platform (at a height of about 23 m), the other on the highest possible beam of the tower stem (about 21 m on the Daniłowicz headframe and about 20 m on the St Kinga headframe) (Fig. 3).

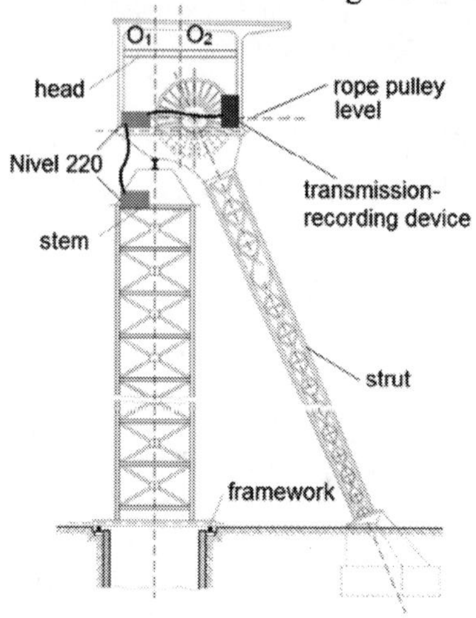

Fig. 3. Scheme of sensor arrangement on a headframes. O_1 – stem axe, O_2 – head axe.

4 Data analysis

The system was launched at the end of October 2016. Due to large changes occurring in the first days of data logging, it was decided to analyse the deviations from November 1st, 2016. Currently, laboratory tests are underway on the time of stabilization of the system depending on the time of transport, the position of sensors in transport and changes in temperature or voltage. This will allow future errors in the interpretation of data to be avoided. One of the biggest problem of the analysis were data writing errors. They were probably caused by temporary power interruptions or overvoltage. The second problem of the analysis was measurement noise caused by diurnal changes in the inclination. Under the operating conditions of the headframes in the Salt Mine Wieliczka, they are conditioned by:

- wind pressure on the structure
- operation of hoisting equipment machines
- insolation of the structure.

Analysis of deflections does not require monitoring of changes in the momentary deflection of the tower, so it was decided within an interval of 68 seconds, hence it was decided to take into account only observations from each day at 4 am (CET). The temperature is then stabilized after a night cooling (this is especially important in the summer). The Wieliczka Salt Mine works in a 2-shift system (6 am to 10 pm), while the service of tourists takes place between 6 am and 7 pm or 9 pm (depending on the season).

Occasional events organized underground can take place in the evening, but usually end around 2 am. At 4 am therefore, the impact of the work of the hoisting machines is not expected.

In order to eliminate errors a program was created in the Python environment to denoise outliers. It was based on the selection of the appropriate data line, its analysis for correctness of data and deletion or saving to the result file. It was also necessary to check the correctness of the temperature mark, because only four characters were recorded, counting a positive or negative sign and a comma, however, the signs determining the ratio to zero were given up first. This means that at a temperature of more than 10 degrees Celsius, the character was not written to the file. For a specific day, a sign like in the previous day was assigned in such cases. The files were then formatted to a compatible version with a browser allowing additional analyses in MS Excel. The data viewer in the LabView system only allows to visualize the changes in inclinometer readings. The data collected in MS Excel made it possible to calculate periodic changes of deflection at individual levels and resltant vectors (total vectors of deflection without adopting a specific value of initial eccentrity vector M_i). Below are presented graphs of periodic changes in the deflection of the Saint Kinga (Fig. 4) and the Daniłowicz headframes (Fig. 6) and resultant vectors of the Saint Kinga shaft declination (Fig. 5) and the Daniłowicz headframe declination (Fig. 7). Sketches of towers with the designation of their own layout and the location of the sensors are shown in Fig. 8.

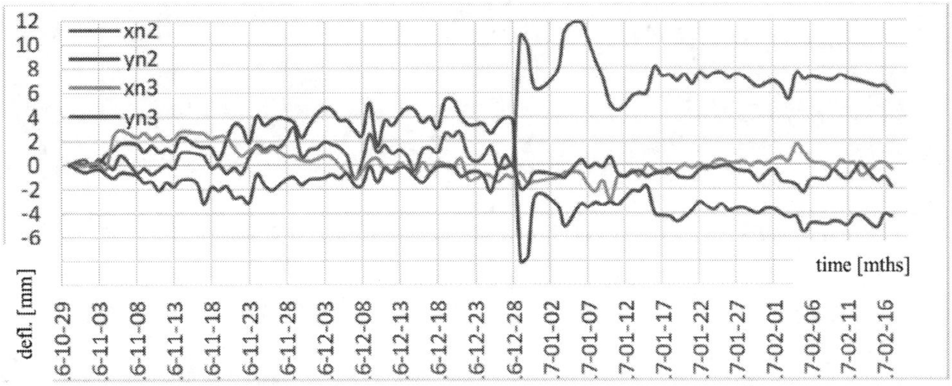

Fig. 4. Periodic deflection changes at the head (N2) and stem (N3) level, St. Kinga shaft.

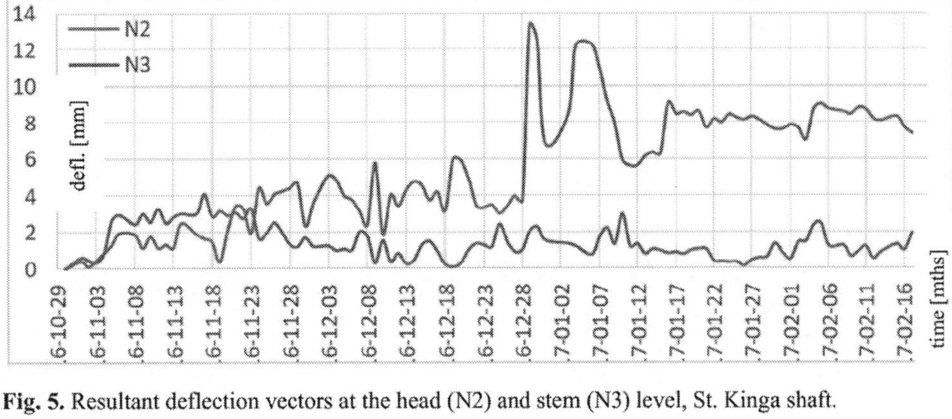

Fig. 5. Resultant deflection vectors at the head (N2) and stem (N3) level, St. Kinga shaft.

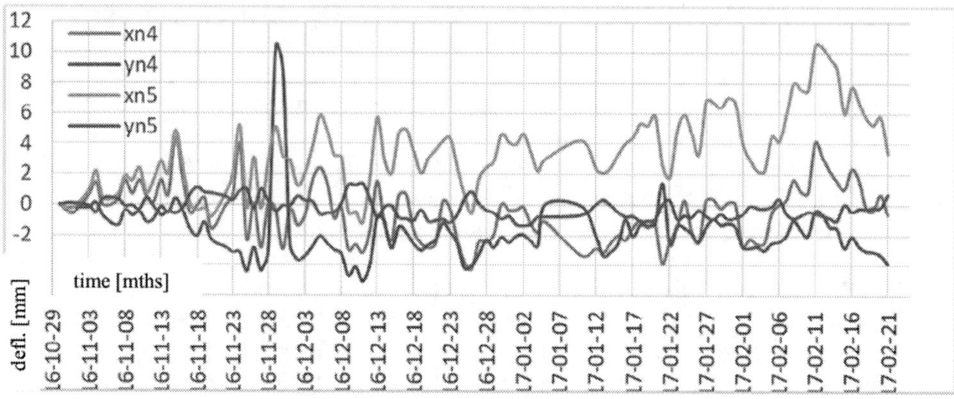

Fig. 6. Periodic deflection changes at the head (N4) and stem (N5) level, Daniłowicz shaft.

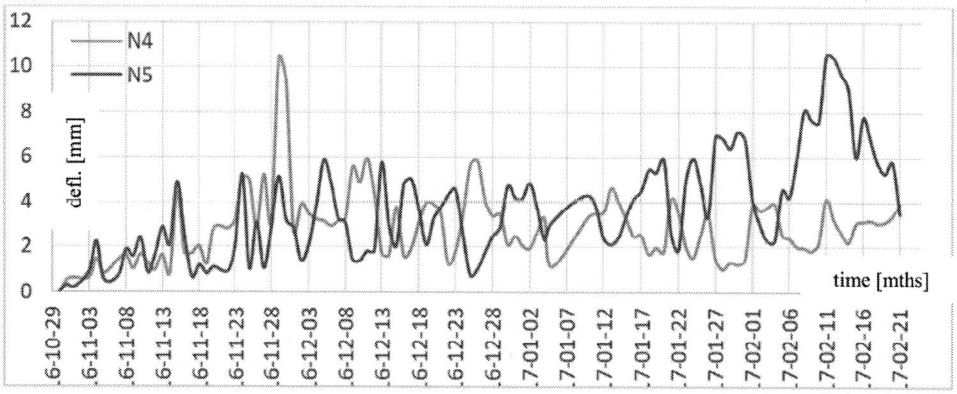

Fig. 7. Resultant deflection vectors at the head (N4) and stem (N5) level, Daniłowicz shaft.

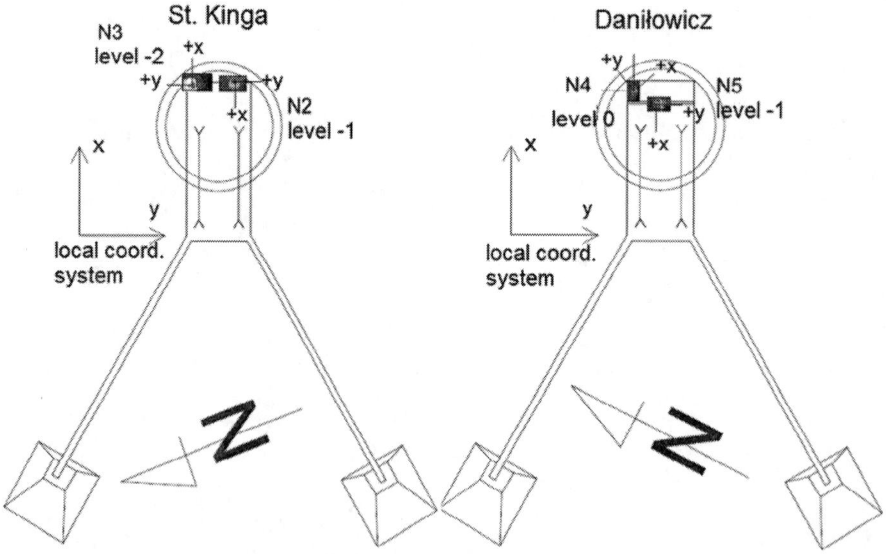

Fig. 8. Sketch of headframes with designation of local coordinate system.

After installing the sensors and the reading system at the end of October 2016 and carrying out the review in mid-November 2016 and the inability to start the data transmission by the wi-fi module, it was decided to remain in the data recording mode on the SD card. This was related to the need to enter the headframe each time in order to collect data. The next time for the system review and data collection was February 21, 2017. The inspection did not show any problems, the transmission devices worked well and the inclinometers were in the range of A or B measurement accuracy [8]. During the next review carried out in June 2017, it was found that the data from the SD cards can not be read and the display with data is not working. Later attempts to recover data in professional companies have confirmed the burning of cards and data loss. The technical service showed that the system was damaged due to overheating caused by significant overvoltage, not necessarily from the power grid (for example as a result of a lightning strike). In both recording devices (on both headframes) all semiconductors have been damaged. The failure limited the possibility of analysis to the period from Nov 1st, 2016 to Feb 21st, 2017. The system resumed its work in January 2018.

5 Summary

The basic problem of the system turned out to be its durability. The system between February and June 2016 crashed, resulting in a loss of data from this period and a long-term service. The defect would be detected faster and less data would be lost if the observations were sent via the wi-fi module. Currently, work is underway to strengthen the transmission system and secure the system against voltage changes. The inclinometers themselves are currently subjected to laboratory tests on the influence of temperature, shocks and the time needed to fully stabilize the stabilization reading after transport and installation.

In the period in which data was correctly recorded on the SD card, two significant incidents were recorded, causing excessive declinations. In the case of a headframe located above the Saint Kinga shaft it was the period from December 28, 2016 to January 8, 2017. From the information provided by the Wieliczka Salt Main, it appears that during this period works related to the exchange of ropes were carried out, which would have caused the displacement of the head itself. A similar event, however unrelated to the setting of ropes, took place on the headframe over the Daniłowicz shaft on November 28-30, 2016. The cause of such inclinometer was not identified and the readings returned to the previous trend after a few days.

These events confirm the previous comments regarding the Regulations (from 2002 [3] and 2017 [2]), in which the axis of the headframe is defined. Studies clearly indicate that the head and the stem of the headframe behave independently of each other, so it is impossible to determine the axis of the entire object. The authors invariably propose the delimitation of the problem of verticality of the guides and the stem of the tower as one element, and the verticality of the rope descending from the rope wheels to the shaft as the second and independently deformation of the headframe as a building object.

This article was developed within the statutory work of AGH US 11.11.150.195 and 15.11.150.405.

References

1. Polish Norm: PN-G-09051, p.14 (1997)
2. Regulation of the Minister of Energy on the detailed requirements for the operation of underground mining facilities (November 23, 2016)
3. Regulation of the Minister of Economy on health and safety, traffic management and specialized fire protection in underground mines (28 June 2002)
4. M. Jabłoński, W. Jaśkowski, T. Lipecki, Telemetry measurements..., SGEM Proc. vol.17 iss.22, p. 577 (2017)
5. T. Lipecki, A comprehensive assessment of geometric objects and shaft using laser scanning, Monography AGH University (2013)
6. W Jaśkowski, Modular system... *Archives of Mining Sciences* vol.55 iss.1 p.81 (2010)
7. M. Jabłoński, W. Jaśkowski, The usage... SGEM Proc. vol.17 iss.22, p. 697 (2017)
8. Leica Geosystems AG.: Nivel 210/Nivel220 User Manual (2007)
9. A. Chrzanowski, J.M. Secord, Tilt Measurement, CRC Press, p. 1 (1999)
10. M.J. Henriques, J.N. Lima, S.B. Oliveira, Measuring Inclinations... Dam and Reservoir Engineering Surveying 5507 p. 1 (2012)
11. C. Inal, C.O. Yigit, M. Yetkin, Monitoring of tall building's... 13th FIG Symposium on Deformation Measurement and Analysis, Lisbon (2008)

Possibilities of surface waters monitoring at mining areas using UAV

Ewa Lisiecka[1], *Barbara* Motyka[2] , *Zbigniew* Motyka[1*], *Łukasz* Pierzchała [1], and *Adam* Szade[1]

[1]Główny Instytut Górnictwa, Katowice, Poland

[2]Pekol, Siemianowice Śl., Poland

Abstract. The selected, remote measurement methods are discussed, useful for determining surface water properties using mobile unmanned aerial platforms (UAV). The possibilities of using this type of solutions in the scope of measuring spatial, physicochemical and biological parameters of both natural and anthropogenic water reservoirs, including flood polders, water-filled pits, settling tanks and mining sinks were analyzed. Methods of remote identification of the process of overgrowing this type of ecosystems with water and coastal plant formations have also been proposed.

Keywords: surface waters, monitoring, UAV

1 Introduction

Among the series of remote measurement methods, useful for determining surface water properties, such as satellite spectrometry methods, there have been opportunities to improve many of them through the use of mobile unmanned aerial platforms (UAV).

As part of the presented work, the possibilities were analyzed of using this type of solutions which can be widely used in the scope of measuring a number of spatial, physicochemical and biological parameters, both natural and anthropogenic water reservoirs, from pond and settler supervision to monitoring of dam reservoirs and watercourses. In addition, the possibility of remote inspection from a short distance of technical condition of hydrotechnical devices and analysis of the structure of the coastal zone.

2 Scanning the spatial geometry of anthropogenic formations

The UAV mobile unmanned platforms were used until now only for terrestrial measurements [1-2].

[*] Corresponding author: motyka@gig.eu

In the first stage of works performed in Główny Instytut Górnictwa (Central Mining Institute), titled "Remote laser systems for determining geometrical parameters of anthropogenic forms of terrestrial surface formation", the methods have been proposed that can also be used to study such structures using laser scanners mounted on UAV platforms, as discussed in more details in previous studies [1-3]. 2D scanning is illustrated on the example of model from Fig. 1. Techniques for further numerical processing of similar fringe images do not differ from those presented in earlier studies [4-6]. However, much simpler effects are obtained with the use of 2D laser scanners that directly determine the distance from the scanning point to a series of points of each line shown in the drawing. Knowing the location of the UAV at any time, to create such a scan we can directly obtain a spatial image of the studied formation in the form of a cloud of points representing it.

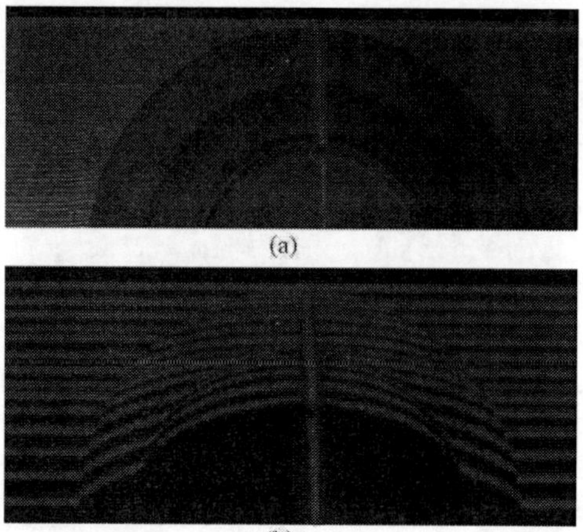
(b)

Fig. 1. Model of anthropogenic terrain formation (a) and 2D scan of the convex part of its spatial geometry for selected scan line 14 positions (b).

Earlier studies discussed the usefulness of laser 2D scanning methods, including remote mining shafts and geometry modeling of mining excavations. With regard to artificial water reservoirs, they will be equally effective for remote inspection from a short distances of technical condition of hydrotechnical devices (weirs, dams, dikes), and analysis of the structure of the embankment zone. While in the case of mine shafts and underground mine workings, the use of drones for laser scanning seems rather distant today and still a fantastic possibility, nothing precludes their use for external inspection of water devices, both with the use of 2D scanners, as well as the traditional photogrammetry from a very short distance.

3 Phytoplankton and coastal plants

Below, the methods of remote identification of the overgrowth process of this type of aquatic ecosystems with phytoplankton and coastal plant formations are also proposed (Fig. 2).
Satellite monitoring of flood areas, physicochemical parameters of smaller and larger water reservoirs, including the quality of their waters, are burdened with many limitations such as: spatial and spectral resolution of satellite scanners, suppression of radiation reflected from the surface of water through the atmosphere of the Earth, or observation angle. For this reason, spectrometric techniques for measuring the quality of water from close range (from a boat or from a bank of water reservoir) are developing, including chlorophyll-a

content, turbidity and other physicochemical parameters such as PRI (Photochemical Reflectance Index), NDVI (Normalized Difference Vegetation Index) and others.

(a) (b)

Fig. 2. Aerial photograph (a) and shore (b) of the artificial water reservoir in Łodygowice, covered with, among others, flowering formations of *Heracleum sp.*

Chlorophyll-a is the basic indicator of surface water quality, illustrating changes in the concentration of phytoplankton in water. Enrichment of aquatic ecosystems with biogenic substances (eutrophication process) results in excessive growth of lower vegetation (in particular algae and cyanobacteria). This process leads to a decline in biodiversity and a reduction in the potential of reservoirs to perform an economic and recreational function [7-8]. Assessment of the concentration of phytoplankton biomass using traditional methods requires direct sampling of water samples in the field and conducting costly and time-consuming analytical methods.

Chlorophyll causes spectral changes in water. It is possible to determine it on the basis of spectral measurements.

The authors of the work carried out research work on the development of a chlorophyll-determination method in inland waters. For this purpose, the measurement of the spectral characteristics of solar radiation reflected from the water surface was used. The device recording the measurement data, in this case, was a monochromator spectrometer with a diffraction grating in the Czernego-Turner system and a photomultiplier as a detector.

Another method used to study the chlorophyll-a content in surface waters, especially in marine and ocean waters, may be the analysis of satellite data.

Inland waters are characterized by greater biological and optical complexity, which considerably hinders the application of spectral methods. Very often, the spectral characteristics of individual phytoplankton and suspension groups overlap, which forces the development of new algorithms that allow obtaining information for given surface water types [9]. The development of one universal calculation formula is therefore virtually impossible and requires in situ comparative testing. The application of the water quality monitoring method using photos from the Landsat system on lakes in Oklahoma allowed for a high correlation with water turbidity. However, there was no significant correlation of the chlorophyll-a content with individual spectral bands and their combinations [10]. High correlation coefficients (0.944) between the actual chlorophyll-a concentration, calculated on the basis of spectral data, were obtained for a lake in New Zealand [11]. The SCyanoHABs system is an example of an effective implementation of the method for assessing the ecological status of lakes in the Mediterranean zone based on satellite images. Due to the low resolution of satellite images used in this method (MERIS system), this monitoring covers only the largest water reservoirs in Spain [12]. The limitations for methods using satellite images for surface water monitoring are also atmospheric conditions.

The persistent cloudiness limits the possibility of satellite observation of the tank's surface. Conducting in situ observations allows for providing data with a higher resolution, without the disturbing influence of atmospheric conditions. To analyze the spectral characteristics of light reflected from the surface of water in situ, both specialized spectral cameras and relatively inexpensive digital cameras can be used [13]. The presence of chlorophyll-a is manifested by an increase in the spectral reflection, in the green and red radiation range, and increased absorption in the area of the blue spectrum. The analysis of information on the intensity of component colors from digital photography in graphic software allows the calculation of coefficients allowing the assessment of chlorophyll concentration in surface waters. Initial research using photographs taken with the help of a digital camera placed on a mobile unmanned flying platform indicates the high potential of this solution in the monitoring of inland surface waters [14]. The possibility of assessing the chlorophyll concentration on the basis of the analysis of the spectral characteristics of light reflected from the water surface is also confirmed by preliminary tests carried out with the aid of the presented measuring apparatus.

The UAV platforms prove to be particularly useful for the remote identification of overgrowth of water ecosystems with hardly accessible shoreline with phytoplankton, surface plants and coastal plant formations such as *Heracleum* sp., *Petasites* sp., *Phragmites* sp., *Lemna* sp., and many others [1,15] (Fig. 3 and Fig. 4).

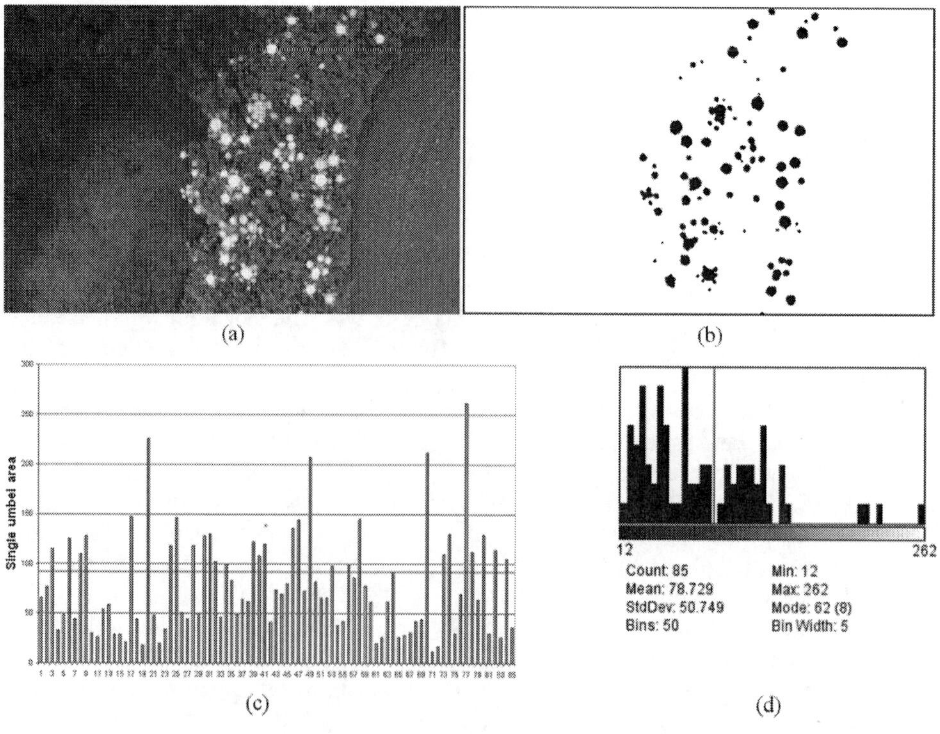

Fig. 3. Identification from the air of the overgrowth area of a small, anthropogenic water ecosystem with the water and coastal plant formations (a), especially from the *Heracleum* genus (b) allows to obtain a quantitative distribution of the canopy sizes of this plant (c) and (d).

 https://doi.org/10.1051/e3sconf/20183602004

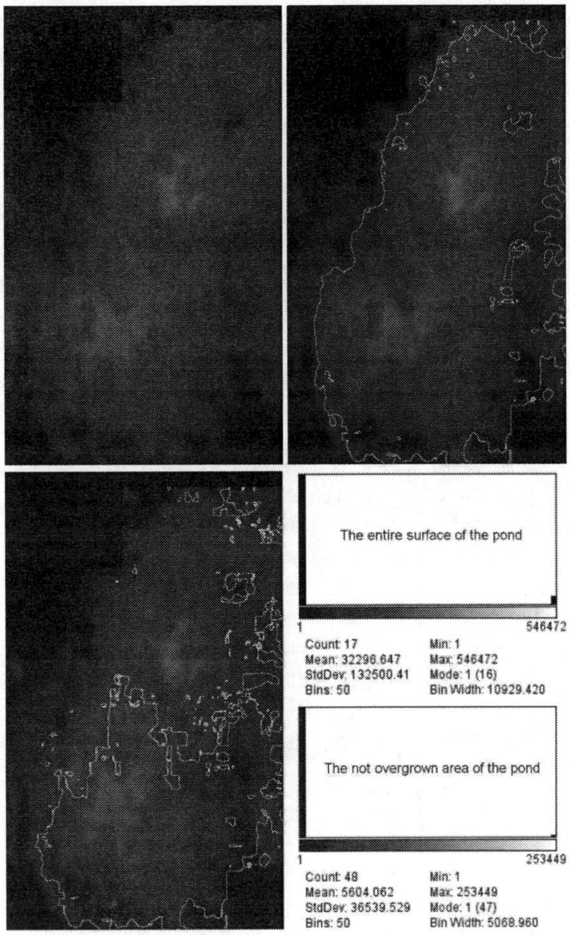

Fig. 4. Identification from the air of the overgrowth area of a small, anthropogenic water ecosystem with surface plants by image processing with the use of ImageJ software packet.

In the case of *Heracleum* sp., from the size distribution of umbel, it is possible, for example, to estimate the number of blooming individuals. In Figure 3, ImageJ software packet was applied to photograph (a) to extract with the use of picture processing negative images of white canopies only. Within the same packet a procedure was used counting number of black pixels for each of 85 black stain. The latter are presented in sequence (numbered according to the counting procedure advance) with the use of Excel (c). Their size distribution (d) was obtained with the use of another ImageJ procedure. The estimated visually number of approximately 31 blooming individuals remains in good accordance with the number of umbels with an area of nearly 90 pixels or more, indicated by red lines in diagrams (c) and (d).

The ImageJ software packet was also used to analyze the area of overgrown part of the adjacent pond which upper part is also visible at the left hand side of *Heracleum* sp. formation from Fig. 3a.

Its image colors were enhanced and details were blurred with the use of procedures of ImageJ to the form of a top left-hand picture from Fig. 4. Another procedures of this packet enabled to delineate areas of different colors corresponding visually overgrown part of the pond only (the bottom left-hand picture from Fig. 4) and its entire area (the top right-hand picture from Fig. 4). The both areas were calculated as a number of pixels (253449 and

546472) contained within them with the use of procedure of ImageJ and the results were presented at the bottom right-hand side diagrams of Fig. 4. The ratio of the area not overgrown to the entire surface of the pond from Fig. 4 is approximately 0.46.

4 Surface water pollution

In the first stage of works performed in GIG in 2016, titled "Monitoring of selected parameters of low emission and location of surface water pollution with the use of drones", a measurement platform for selected low emission parameters was set up for drones in the form of a hybrid optoelectronic meter for CO, CO2 and low emission dust fractions. The use of drone as a mobile carrier for the measuring platform plays a key role here. Combustion (co-firing with coal) of plastics (PVC, polyurethanes, PET) results in the formation of very toxic compounds, such as hydrogen chloride, carcinogenic dioxins and furans, formaldehyde and others, which may then flow from the air to surface waters as a result of precipitation. Therefore, it is important to determine such components already in the air in the vicinity of or over open water reservoirs. For example, the possibility of measuring the distinctive characteristic component, such as HCl, as well as the possibility of collecting gas samples for further analytical laboratory tests based on procedures based on current regulations and standards, may be important here. The device for remote laser determination from the UAV deck of physical-geometrical parameters of anthropogenic forms of terrain surface formation, developed in GIG, may also be useful for mapping water reservoirs and air pollution above them [3]. The second stage of the project involves the construction of drone equipment with an infrared and spectral camera for the location of discharge of pollutants into waters, as well as the extension of applications based on GIS.

Summary

The possibilities of using photogrammetry with the use of UAV for mapping the coastal zone of specific plant cover, as well as overgrowing of open water reservoirs were shown.

Another area of application for drones may be the construction of measuring platforms that will allow the assessment of surface water quality and the identification of pollution discharge points. The next stage in the development of remote methods for measuring surface water quality may be the placement of measurement devices recording spectral reflection from the water surface, in interesting spectral ranges or for selected wavelengths, on a mobile unmanned aerial platform. Such a solution will allow adjusting the parameters of the remote sensing device to the conditions in which the measurements are carried out and to carry out measurements in situ directly over a specific water reservoir.

References

1. Motyka Z., JCEEA, XXXIV, 64, 2/II/17 (2017).
2. Motyka Z., In: Innovative development of the mining industry (Kryvyi Rih National University, Kryvyi Rih, 2016).
3. Motyka Z., P.417785, BUP 4 (2017).
4. Motyka Z., Proc. SPIE, 5229, 10 (2003).
5. Motyka Z., Wiadomości Górnicze 5 (2005).
6. Motyka Z., Passia H., AIP Conf. Proc., 1457/1 (2012).
7. Scheffer, M.Ch.H., Ecology of Shallow Lakes (London, 1998).

8. Pierzchała, Ł., Sierka, E., Trząski, L., Bondaruk, J., Czuber, B., AEER 14, 1 (2016).
9. Kruk, M., Rychter, A., Mroz, M., Zalew Wiślany – środowisko przyrodnicze oraz nowoczesne metody jego badania na przykładzie projektu Visla (PWSZ, Elbląg, 2011).
10. Barrett, D. Clay; Frazier, Amy E., Water, 8.6, 257 (2016).
11. Allan, M.G.; Hamilton, D.P.; Hicks, B.J.; Brabyn, L., Int. J. Remote Sens. 32 (2011).
12. Gómez, J. A. D., Alonso, C. A., García, A. A., EMA, 181, 1-4, (2011).
13. Goddijn-Murphy, L., Dailloux, D., White, M., Bowers, D., Sensors, 9, 7 (2009).
14. Cândido, A. K. A. A., Paranhos Filho, A. C., Haupenthal, M. R., da Silva, N. M., de Sousa Correa, J., & Ribeiro, M. L., WASP, 227, 7 (2016).
15. Motyka B., Motyka Z., In: Zwalczanie inwazyjnych gatunków roślin obcego pochodzenia – dobre i złe doświadczenia (Stowarzyszenie „Człowiek i Przyroda", Krzywe, 2015).

The use of the DInSAR method in the monitoring of road damage caused by mining activities

Radosław Murdzek[1], *Hubert* Malik[1] and *Andrzej* Leśniak[1]

[1]AGH University of Science and Technology, Faculty of Geology, Geophysics and Environmental

Abstract. This paper reviews existing remote sensing methods of road damage detection and demonstrates the possibility of using DInSAR (Differential Interferometry SAR) method to identify endangered road sections. In this study two radar images collected by Sentinel-1 satellite have been used. Images were acquired with 24 days interval in 2015. The analysis allowed to estimate the scale of the post-mining deformation that occurred in Upper Silesia and to indicate areas where road infrastructure is particularly vulnerable to damage.

Keywords: DInSAR, road damage, Sentinel-1

1 Introduction

One of the biggest challenges in road infrastructure management is early detection of its damages and taking actions to reduce repairing costs. Many studies shows that early detection of damages increase the road quality life and reduce the total maintenance cost. Therefore, there is a need for road monitoring methods that facilitate the detection of the road damage.

Roads and railways are very sensitive for ground deformation. In the paper [1] authors described destructive effects of subsidence in case of linear, large infrastructure objects (like railway and highway). Authors presented parameters responsible for horizontal deformation, inclination and curvature as a reason of problems with infrastructure and even its destruction. Among other threats, changing the slope of the terrain can lead to slide off slowly moving vehicles inside the curve or slide off outside the curve of the track when driving at the maximum speed for which the radius of movement was designed. Horizontal deformation of the ground surface may lead to loosening of the ground of embankment and lowering its bearing capacity and to cracks in the pavement structure. In the case of vertical displacements, there may also be changes in water relations within the road and its neighbourhood. It have to be note that some buildings are also part of the road infrastructure. Vertical dislocations may cause serious damage in the form of additional compression, tensile, torsional stress, etc. This can significantly limit or even prevent from use of communication routes subjected to such processes. Hence the monitoring of road damage can be extremely important both in terms of safety and minimization of repair costs.

Many authors indicate the large potential for the use of satellite remote sensing in detecting road infrastructure damage. One of the remote sensing system is Synthetic Aperture Radar (SAR). SAR system is most often mounted on satellite platform. It emits microwaves towards Earth surface. Back-scattered rays are collected on SAR antenna. It gives information about signal strength and its phase for each resolution cell in SAR imagery [2]. The aim of this work is to review existing SAR methods of road damage detection as well as to demonstrate the possibility of using Differential Interferometry SAR (DInSAR) method.

2 Road damage detection using remote sensing data

Remote sensing methods can be used to detect road damage. The most commonly used sensors to accomplish that task are: multispectral camera, hyperspectral camera, Synthetic Aperture Radar (SAR), LIDAR and Terrestial Laser Scanning. Some methods work in a direct way, indicating places where the road surface is cracked while others indirectly - indicating potentially endangered areas (for example where the ground was unstable), which could lead to road damage. The detection methods presented here concerns paved roads built of such materials as asphalt or concrete cement.

2.1 Multispectral reflectance analysis

One of the most commonly used remote sensing methods of investigating road damage is spectral reflectance analysis. Paved road structural damages like cracking result in high reflectance detected by multispectral sensor. On the other side, undamaged, good quality paved road shows lower reflectance values.

Wei et al. in [3] analyze the spectral properties of asphalt in order to distinguish an old and deteriorated road from a new-built road. Authors conclude that the minimum reflectance for each type of road is near 350 nm wavelength and it rises with wavelength used in sensing system. The most noticeable differences of the observed reflectivity for the new and old roads are visible in the SWIR regions – 1750 nm wavelength. It is also worth to note, that different types of asphalt reflect radiation in various ways, so only roads of the same type of material can be compared.

In [4] authors use airborne MIVIS (Multispectral Infrared Visible Imaging Spectrometer) sensor data. The results of this study demonstrate the utility of airborne remote sensing for discriminating roads which requires a maintenance work. When the road paving asphalt is made of a mixture of limestone granules and bitumen, then surface defects are related to the decrease of the oily components of the bitumen. This causes percentage increase of limestone granules fraction in an asphalt. Such phenomenon enables to distinguish deteriorated surface from a good quality one by retrieving the emissivity spectral features of asphalt in 8-13μm TIR range.

Another work points out [5], that not only asphalt has higher reflectance values as it degrades, but also becomes less uniform in texture on multispectral images. The calculation of asphalt pavement statistics in the image indicated that road damages leads to broadening of the data range on image and variance and entropy increase.

2.2 SAR backscattering analysis

The usefulness of SAR data in the study of road quality has been confirmed by many authors. Especially the robust method is the backscatter signal analysis registered on an image as an amplitude.

Suanpaga and Yoshikazu [6] proposed an approach to determine the level of highway riding service using SAR data. The aim of this work was to investigate the influence of backscattering values of Phase Array SAR (PALSAR, L-band). Performed analysis showed that a poor condition of the road was manifested in the image by higher backscattering value for copolarization (HH or VV polarization). Authors concluded that the most robust variable for developing riding quality evaluation was the HH backscattering value. The accuracy of presented approach was 87.00%.

Another study [7] exploits SAR data acquired at X-band frequencies. Authors compare high-resolution Cosmo-SkyMed data with road roughness data obtained with ground surveying methods. Research work shows significant correlation of SAR image radar cross section with pavement roughness. At X-band radar brightness generally increases when pavement roughness worsens. The overall accuracy obtained with this method was 92%.

2.3 Persistent Scatterer Interferometry

One of the satellite methods of road monitoring is Persistent Scatterer Interferometry (PSI). This method does not provide direct information on road surface damage. Its result is a vertical displacement map of persistent scatterers points. Such points may be buildings, rock outcrops or elements of road infrastructure. So this method enables to monitor the deformations of roads, their infrastructures and, surroundings. Its technical assumptions were described in [8].

An example of the use of the PSI method has been presented in [9]. The PSI analysis was performed using high-resolution TerraSAR-X data. Authors used 28 SAR images in order to derive vertical displacement map of pints in Barcelona metropolitan area. The analysis enabled to detect fragments of a new part of C-31 highway that is threatened with ground subsidence. Subsidence phenomenon is caused here by recent construction of this part of the highway, not by underground mining excavation.

Another example of the use of PSI technique is described in [10]. The purpose of this work was to detect active landslides in the Carpathian region based on ERS-1/2 satellite data obtained before the year 2000. Analysis revealed a linear deformation of the area along the Osielec-Jordanów road of 3.8 mm/year. In 2008, this road required renovation due to the damage caused by an active landslide. However authors indicates that analysis of landslide deformation using PSI in Carpathians is possible, but very difficult due to, among others, rough topography and dense vegetation cover.

3 Differential Interferometry SAR method

Another indirect method of detecting road damage as a result of mining activities is the detection of subsidence troughs that occurred under road infrastructure. To obtain vertical ground subsidence information SAR images can be processed with interferometric methods. One of them is Differential Interferometry SAR (DInSAR). To compute differential interferogram, at least two same-area SAR images are required. One of them is a reference image (master). Comparing master image with the second image (slave) it is possible to compute vertical ground deformations that occurred between time of slave and master images acquisition [11,12]. The inclusion of another SAR images into analysis enables to perform ground deformations analysis through time. DInSAR method accuracy is about 1 cm in vertical direction. This method is particularly suitable to detect post-mining subsidence phenomenon. On differential interferogram ground deformations are represented as fringes of wrapped phase (values between $-\pi$ and π). Then fringes can be converted into ground displacement values in the satellite Line of sight (LOS) direction. One interferometric fringe (in case of DInSAR analysis) is equal to the half of the SAR system wavelength.

Vertical displacements information can be used to identify sections of roads and railway tracks that have been (or may be) damaged due to mining activities. Single scene of radar imagery covers large area (even over 15000 km^2) which makes it possible to analyze all important roads in the city with a single analysis.

3.1 Exploited dataset

Ground subsidence can be caused by natural phenomena, but it can also be an effect of anthropological processes. Most common reason of subsidence caused by human activity is underground coal mining. In Poland the most intensive coal excavation activities are located in Upper Silesian Coal Basin (USCB). Even on areas where mining activity was stopped subsidence can occur [13].

DinSAR analysis was performed for Bytom city that is located within USCB. This city was selected for analysis due to the mining activity that has been conducted there for many years. Transport routes of regional significance run through Bytom. Many of them have been damaged as a result of underground mining excavation. Analysed city covers area of 69,5 km^2.

In this paper Sentinel-1A (5,5 cm wavelength) images acquired in 29.11.2015 and 23.12.2015 were used. 24 days time interval was chosen to achieve good coherence of radar images. High coherence made it possible to obtain results with a precision of up to 1 cm. Satellite data were processed in SNAP software provided by European Space Agency.

3.2 Results

First step of the analysis was to compute differential interferogram representing the surface deformations that occurred between 29.11.2015 and 23.12.2015. Quality of the interferogram can be measured with coherence values. Coherence is determined by the degree of compliance of the SAR signals phase – the higher the value, the more accurate the interpretation of the interferogram will be. Areas where coherence value is below 0.3 are not suitable for DinSAR analysis. The average coherence value of used images in Bytom city is 0.47.

Then interferogram was converted to ground subsidence map (fig.1). One subsidence trough in the central part of Bytom was detected. Depth of this trough was about 6 cm. The centre of detected trough is localized at 18°52'10"E, 50°21'24"N.

4 Discussion

The urban area where the greatest threat related to land displacement was detected is the central part of the city of Bytom (Karb district). It can be noticed that a significant part of the subsidence trough is located under the motorway interchange. This creates a risk of damaging the road infrastructure. This trough reaches 6 cm depth and covers the part of DK94 and DK88 roads, as well as the Family Garden Parks "Rekreacja", residential buildings between "Warszawska" street and "Celna" streets. The area of the trough is 0.95 km^2.

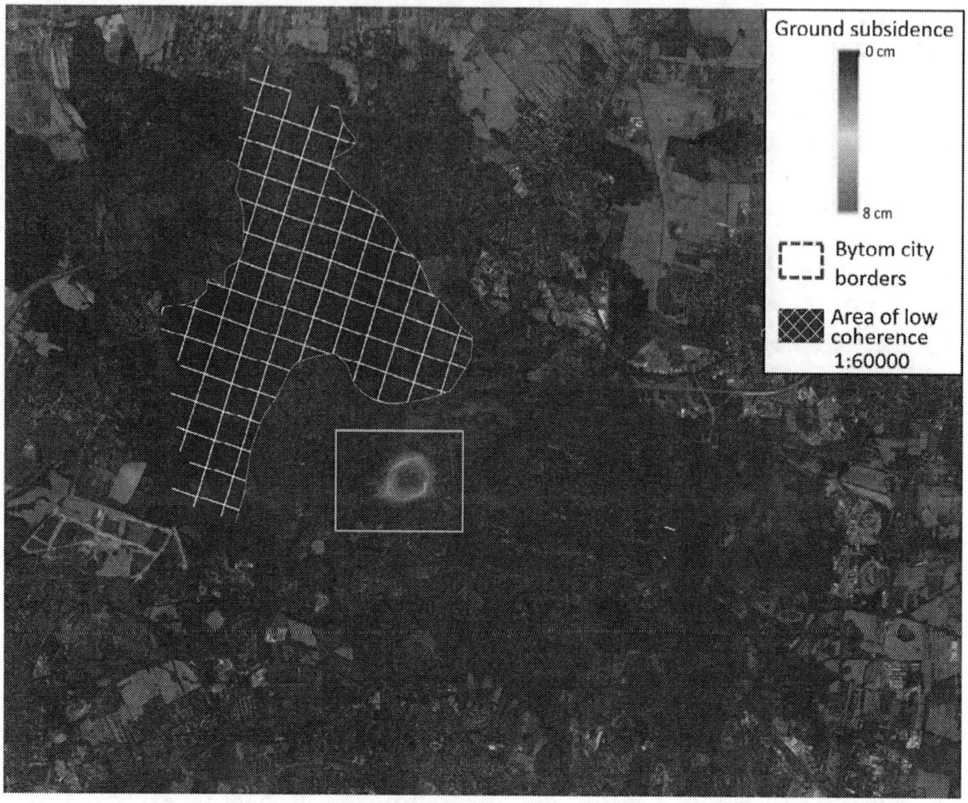

Fig.1. Map of ground subsidence that occurred between 29.11.2015 – 23.12.2015. One subsidence trough was detected in the central part of Bytom city. Area marked in white is the area of low coherence and was not included in analysis.

Area under DK94 route has collapsed on the section of about 1200 m, while under DK88 route on the section of about 1000 m. In both cases, the maximum subsidence exceeded 5 cm. It should be emphasized, that the detected terrain deformations occurred within only 24 days (between November 29, 2015 and December 23, 2015). Mining damage caused by mining operations conducted under the Karb district in Bytom has been registered for many years. Negative effects of coal mining have damaged, among others, water supply systems in the district, viaduct on Wrocławska street, or a nearby gas station [14]. In addition, in June 2016, the restaurant "McDonald's" located at Konstytucji street was closed. The reason for the decision was the poor technical condition of the building and high maintenance costs related to the growing costs of removing mining damage. Perhaps if the monitoring of the stability of this area was carried out earlier, the decision to build a restaurant in this place would not have been made and the DK88 and DK94 routes would be planned in a different way to avoid its destruction.

5 Conclusions

The results presented above confirm the usefulness of DInSAR method in detecting sections of roads threatened by damage associated with underground mining activities. This method does not show road surface damage but indicates the road sections where these damage occurred or may occur in the future. Such approach allows to identify the endangered sections of not only large road sections , but also the smaller ones due to the indication of the whole area where ground subsidence was detected.

The differential radar interferometry method is well suited for monitoring of road damage caused by mining activities because of:

- single radar scene covers large area, which enables effective monitoring of the entire road network,
- results obtained for the entire area, not only in individual measurement points (contrary to PSI method),
- measurement accuracy of 1 cm vertically (for coherent areas),
- no need to conduct field tests,
- the type of asphalt covering the road does not limit the applicability of the method,
- prior knowledge about reflectance curve of undamaged road is not necessary (contrary to multispectral reflectance analysis method),

Despite many advantages, there are also some limitations of DInSAR method. First of all measurements are possible only for coherent areas. Moreover DInSAR does not allow the detection of horizontal displacements or discontinuous deformations. Nevertheless, the DInSAR method used together with other remote sensing methods can be a very powerful tool for detecting road damage.

This work was financed by the statutory research funds of the Department of Geoinformatics and Applied Computer Science AGH No. 11.11.140.613

References

1. K. Tajduś, A. Misa, Przegląd Górniczy 5, 70, 39-47 (2014)
2. J. Strzelczyk, S. Porzycka-Strzelczyk, IEEE Geoscience and Remote Sensing, 4, 783–787 (2014)
3. J. Wei, G. Zhou, Z. Zheng, ASPRS 2009 Annual Conference Baltimore, Maryland March 9-13 (2009)
4. S. Pascucci, C. Bassani, A. Palombo, M. Poscolieri, R. Cavalli, Sensors (Basel), 8(2), 1278–1296 (2008)
5. W. Emery, A. Yerasi, N. Longbotham, F. Pacifici, Assessing paved road surface condition with high-resolution satellite imagery, IGARSS, Presentation (2014)
6. W. Suanpaga, K. Yoshikazu, Remote Sens., 2, 2531-2546 (2010)
7. F. Meyer, A. Olaniyi, E. Hoppe, IGARSS, pp. 1558-1561 (2017)
8. P. Rosen, S. Hensley, I. Joughin, F. Li, S. Madsen, E. Rodríguez, R. Goldstein, Proc. of the IEEE, 88 (3), pp. 333-382 (2008)
9. M. Crosetto, O. Monserrat, G. Luzi, M. Cuevas-González, N. Devanthéry, Radar-based remote sensing monitoring of roads, SIRWEC (2014)
10. Z. Perski, T. Wojciechowski, A. Borkowski, Acta Geodyn. Geomater., 7, 3 (159), 1–7 (2010)
11. R. Gupta, Remote Sensing Geology. Springer-Verlag Berlin, Heidenberg (2003)
12. A. Ferretti, A. Monti-Guarnieri, C. Prati, F. Rocca: InSAR Principles: Guidelines for SAR Interferometry Processing and Interpretation, ESA Publications, TM-19, (2007)
13. E. Pilecka, D. Szwarkowski, An application of the ground laser scanning to recognise terrain surface deformation over a shallowly located underground excavation, E3S Web of Conferences, Vol. 24, 1-8, (2017)
14. I. Kantor-Pietraga, R. Machowski, Przemiany przestrzenne oraz społeczne Bytomia i jego centrum, Prace Wydziału Nauk o Ziemi Uniwersytetu Śląskiego nr 75 (2012)

Ground subsidence information as a valuable layer in GIS analysis

Radosław Murdzek[1], *Hubert* Malik[1], and *Andrzej* Leśniak[1]

[1]AGH University of Science and Technology, Faculty of Geology, Geophysics and Environmental Protection, Cracow, Poland

Abstract. Among the technologies used to improve functioning of local governments the geographic information systems (GIS) are widely used. GIS tools allow to simultaneously integrate spatial data resources, analyse them, process and use them to make strategic decisions. Nowadays GIS analysis is widely used in spatial planning or environmental protection. In these applications a number of spatial information are utilized, but rarely it is an information about environmental hazards. This paper includes information about ground subsidence that occurred in USCB mining area into GIS analysis. Monitoring of this phenomenon can be carried out using the radar differential interferometry (DInSAR) method.

Keywords: DInSAR, GIS, underground mining

1 Introduction

In recent years Geographic Information System (ang. GIS) has become an indispensable tool in the process of spatial data analysis. GIS currently has a very wide range of functionalities: creating, acquiring, integrating, processing, visualizing, analysing, modeling and archiving information about the Earth's surface and phenomena that occur on it. These functionalities are used to combine information placed in space and in time to better understand the studied phenomena and to support the decision-making process. GIS analysis finds practical applications in many areas, including spatial planning, creating maps of environmental hazards occurrence, choosing the optimal place for investment, infrastructure management, mining activity planning, etc. Due to the fact that the GIS analysis results are used to make key decisions, they must meet the high users requirements. It is important to use only reliable data for analysis. The more relevant and up-to-date information layers built on the actual data will be used, the more accurate the results will be. This article presents the possibility of using vertical ground deformation information as one of the input data layers in the GIS analysis. The aim of this work is to demonstrate the need of the inclusion of ground subsidence information in spatial analysis, especially for local government units. Radar data is currently available at a very low cost, and the information they provide could be crucial to make certain planning decisions.

In Poland ground subsidence phenomenon associated with underground hard coal mining is a common problem in the Upper Silesia region. Within the Upper Silesian Coal Basin (USCB), 133 coal deposits are documented and according to the Polish Geological Institute, there were still 28 mines operating in this region in 2014 [1]. In the Environmental Protection Programme for the Silesian Voivodeship it was found that the maximum value of ground subsidence reaches even 1 meter per year [2]. Most often, these areas are heavily urbanised and the appearing deformations may be dangerous to the existing building infrastructure [7]. However, even a noticeably smaller terrain subsidence can cause mining damage. In the USCB area, this phenomenon leads, among others, to changes in hydrographic conditions, flooding, formation of marshlands, buildings, roads and other infrastructure facilities damages. Therefore, including ground subsidence information in GIS analysis performed in the Upper Silesia region will lead to improvement of decision-making processes.

To study vertical terrain displacements on large areas satellite radar imagery can be used [3, 4, 5, 6]. The method of processing radar data to obtain vertical displacements information is Differential Interferometry SAR (DInSAR). The guidelines of the method and its case studies have been described by many authors [8,9]. DInSAR method consists of generating a differential interferogram which is an image of the phase difference between two radar images acquired at a certain time interval. The phase difference is proportional to the surface deformation that occurred in the direction of satellite line of sight (LOS). Precision of measurements obtained with this method depends on many factors, the most important of which are: wavelength of signal emitted by the SAR system, spatial and temporal resolution of the images, coherence of two SAR images. As for the DInSAR method it is assessed that the average precision of the vertical displacement measurements reaches 1 cm [10].

2 Characteristics of used data

The presented ground subsidence analysis was carried out in the Upper Silesian Coal Basin area. Analysed area covers approximately 2600 km^2 and it was chosen because of the exploitation of hard coal conducted there on a large scale. Underground coal mining leads to the formation of many subsidence troughs of different range and depth.

In this work, the analysis of vertical surface displacements has been carried out using DInSAR method. For this purpose, two radar images acquired by the Sentinel-1A satellite (C band SAR - 5.405 GHz) at 29.11.2015 and 23.12.2015 were used. Radar images were acquired in Interferometric Wide Swath mode [11].

Radar images were used to obtain a differential interferogram representing the surface deformations that occurred between 29.11.2015 and 23.12.2015. The main parameter that can be used to estimate the quality of the interferogram is coherence. Coherence is determined by the degree of compliance of the SAR signals phase – the higher the value, the more accurate the interpretation of the interferogram will be. In the case of low coherence, interpretation is practically impossible [12]. Fig.1 shows coherent areas (areas where coherence value is greater than 0,3). The average coherence value of used images is 0.36. High coherence values occur in urban and agricultural areas – especially central part of GZW where there are large urban agglomerations. On the other side, low coherence values can be observed on vegetation areas and lakes. Performing DInSAR analysis for such areas is highly limited.

https://doi.org/10.1051/e3sconf/20183602006

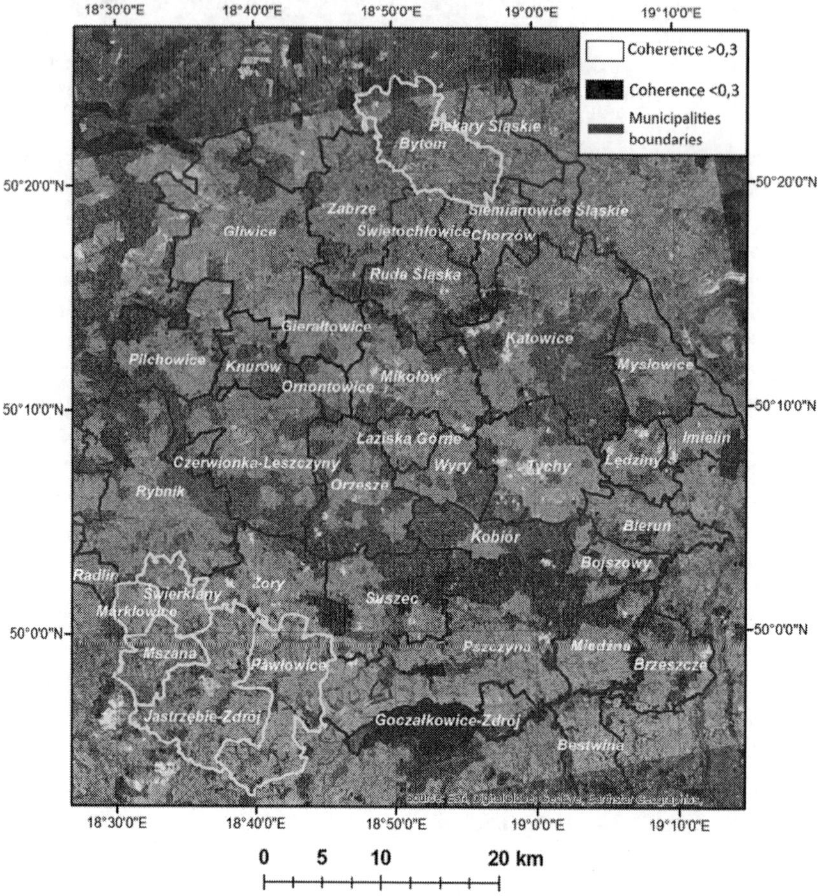

Fig.1. Map of coherent areas overlayed on orthophotomap. Vegetated areas have low coherence values, while urban and agricultural areas - high.

3 Results

As result of the DInSAR analysis, the differential interferogram was generated. The fringes visible on the interferogram could be used to estimate ground subsidence that occurred during the period 29.11.2015 - 23.12.2015, due to the phase differences are proportional to the terrain deformation. Each full range values change on the interferogram fringes represents the ground displacement in the line of sight direction by half the wavelength used by the SAR system (for Sentinel-1 data – 2,75 cm). The obtained results were processed to represent deformations in the vertical direction (Fig.2). In the analysed period 45 subsidence troughs with depths of 1 to 10 cm were identified in the USCB area. The largest subsidence occurred in the municipality of Pawłowice (subsidence trough depth - 10 cm, center of the trough - 49°58'N 18°43'E), Marklowice (subsidence trough depth - 9 cm, center of the trough - 50°02'N 18°31'E) and Gierałtowice (7,5 cm, 50°12'N 18°46'E). All subsidence troughs were detected on hard coal deposits areas, so they can be linked with mining activities.

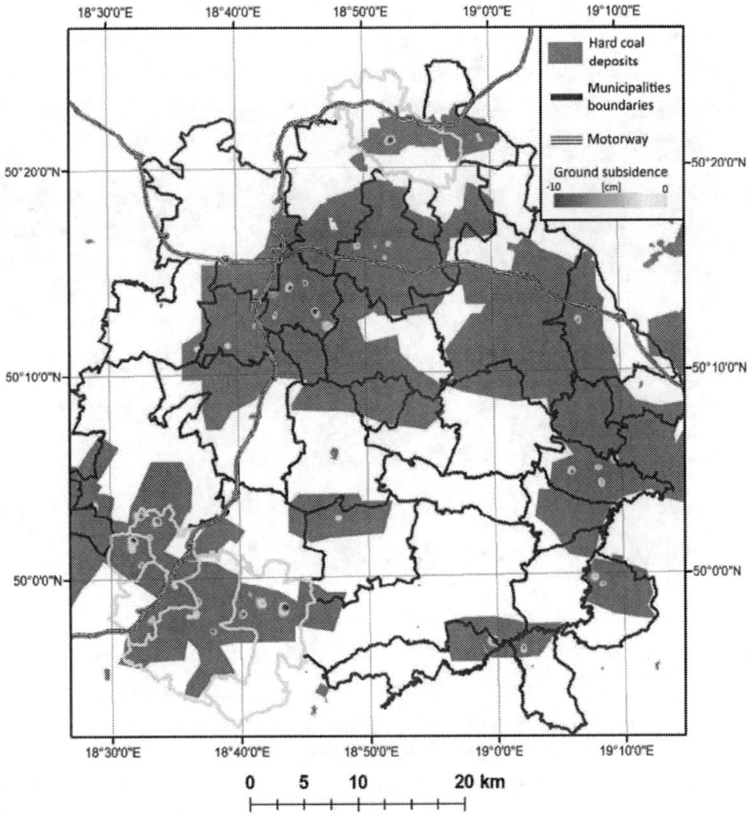

Fig.2. Map of ground subsidence that occurred between 29.11.2015 – 23.12.2015. Subsidence troughs were detected only on hard coal deposits areas.

4 Examples of ground subsidence information use in GIS analysis

4.1 Choosing the optimal location for investment

Geographical data are widely used in investment decisions making process. Appropriate compilation and visualization of this type of data is an important tool both at the stage of investment planning and its implementation. There are many factors that decide which investment plot will be optimal. Most important of them are: utilities infrastructure, intended use in the spatial development plan, accessibility of the transport network, geological and hydrological conditions, demography of the area, slope of the area, insolation, etc. Almost any kind of information about the property can prove invaluable in the context of the investment decision-making. It is therefore important to use a large collection of relevant data.

Historical information about the ground stability of the area, which in the target of the investment, will allow to better assess whether the given plot is risk and effort suitable. In addition, ongoing terrain subsidence monitoring during the implementation of the investment enables quick recognition and undertaking of activities minimizing the

 https://doi.org/10.1051/e3sconf/20183602006

occurrence of damage to objects during their construction. However, it should be noted that the DInSAR method provides only information about subsidence that occurred in the past and does not provide any information whether this phenomenon will occur in the future.

Fig.3. Mean values of ground subsidence within individual plots (29.11.2015 – 23.12.2015). Plots indicated on red are not suitable for investment.

In order to properly interpret the results, 5 thematic layers (orthophotomap, buildings map, roads and railway tracks map, boundaries of plots, ground subsidence information) were put together in order to indicate plots that are not suitable for investment (Fig.3). GIS analysis enabled to identify specific plots, buildings and fragments of roads and railway tracks under which the ground is unstable and the risk of mining damage occurrence is high (red colour). It is not recommended to build new buildings in the area located close to the intersection of two main roads, as the impact of underground mining is still high there.

4.2 Designing highway routes

Roads and railway tracks are susceptible to the destructive effects of ground deformation. These deformations are often the cause of cracking and damage of asphalt surfaces. This can significantly reduce or even prevent the use of routes. Vertical surface displacements information can be used to plan the course of new roads and railway tracks. Individual scenes of radar image cover large areas (even over 15 000 km^2), which allows to detect unstable areas in a wide spatial context. When planning a new road, it is necessary to take into consideration existing roads, railway tracks, building, land development, and many other information. In areas threatened by land subsidence it is also necessary to include this information in analysis. The largest threat to infrastructure was detected in the municipalities of Marklowice, Pawłowice and Świerklany (Fig.4). To predict the further development of this threat, information about the planned mining activity should be included in analysis. However, information about the current and past subsidence allows to state that the area is still unstable and exclude them while planning new routes.

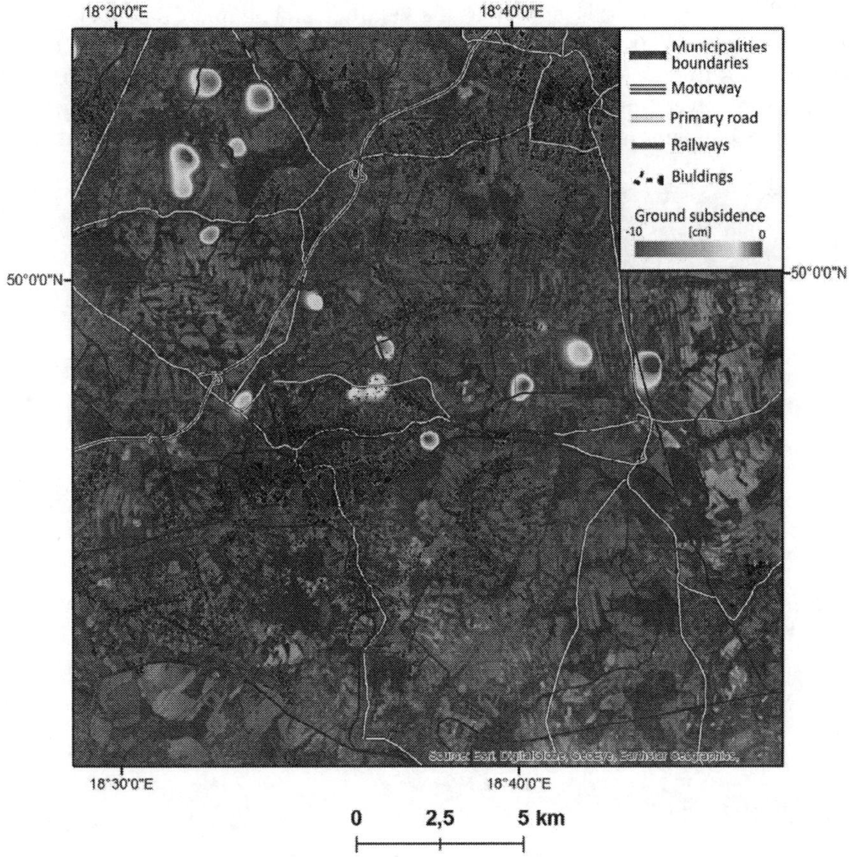

Fig. 4. Map of ground subsidence that occurred between 29.11.2015 – 23.12.2015. In many cases ground subsidence occurs in areas where roads and railways pass through. Such situations should be avoided in road planning.#

4.3 Assessment of mining activity impact on infrastructure damage

Mining exploitation is often carried out under buildings and urban infrastructure. Continuous ground deformations caused by exploitation lead to infrastructure damage on the surface. Mining and construction analysis does not always allow to clearly determine whether the damage was caused by underground mining activity or other factors (such as construction defects of the building, surface vibrations). It is often stated in advance that if the facility is located in a mining area, then mining activity is the cause of the damage. On the other hand, the ground subsidence measurements is performed on a regular basis only by representatives of the mines and they are the owners of these results. In addition, the reliability of these measurements is often questioned by the owner of the damaged object. The lack of other alternative sources of information about ground subsidence causing mining damage is a significant problem in claiming for loss compensation. Satellite data is in this matter a reliable and independent source of information on terrain deformations. It can therefore be used in the process of assessing the cause of surface infrastructure damage.

Fig. 5 shows the following information layers combined: ground subsidence isolines, roads and buildings. The depth of the detected trough reaches 6 cm. The trough covers the part of DK94 and DK88 roads, as well as the Family Garden Parks "Rekreacja", residential buildings between "Warszawska" and "Celna" streets. The area of the trough is 0.95 km^2. In this area infrastructure damages have been observed for many years. The combination of several thematic layers (plots map, roads, buildings) can help to link the damages occurring on individual buildings and roads with the subsidence phenomenon caused by mining activities.

Fig. 5. Map of ground subsidence isolines that occurred between 29.11.2015 - 23.12.2015 in Bytom. Buildings located in the center of the subsidence trough were damaged as a result of underground exploitation in this area.#

4.4 Spatial planning

The local spatial development plan is the basic document at the local level containing urban planning guidelines for the municipality. This document specifies the intended land use of areas. In order to prepare this document, natural hazards, such as flood occurrence, are taken into account. This prohibits constructions of building on endangered areas. Incorporation of ground subsidence information into analysis would complement the analysis with another thematic layer concerning threats. Indication of the areas on which the subsidence troughs still occurs will enable more conscious determination of areas that should be excluded from development. In addition, such information will be an additional source in assigning appropriate mining category to the given area.

The areas indicated on Fig.2 as being threatened with subsidence should be used for agricultural production or recreation. Since these areas are still unstable, they should not be designated for development, and for the construction of large objects. Possibly threatened areas should be designated for reclamation to minimize and mitigate the environmental effects of mining activity.

5 Conclusions

The use of many layers of reliable information in the GIS analysis will allow to minimize the risk associated with making inappropriate planning decisions. Such analysis is cheap, fast and accurate. Analysis based on complete data is crucial especially for local government units. Ground subsidence information can be an important element in GIS

analysis on areas where mining activities are performed or were performed in past. Including this layer in analysis will help to make better decisions in such aspects as:
- security – by excluding areas threatened with ground subsidence from planning new buildings, or in case of detecting the subsidence threat take an appropriate action aimed at infrastructure securing,
- economical – minimization of the costs of mining damage removing by selecting the appropriate area for investment,
- determining the cause and effect relationship between the mining activity and the resulting damage.

Detection of the areas where ground subsidence occurred is sufficient to conclude that these areas are still unstable. This threat must be taken into account in spatial planning.

This work was financed by the statutory research funds of the Department of Geoinformatics and Applied Computer Science AGH No. 11.11.140.613

References

1. Bilans zasobów złóż kopalin w Polsce wg stanu na 31.12.2013 r., Panśtwowy Instytut Geologiczny (2014)
2. Program Ochrony Środowiska dla Województwa Śląskiego do roku 2019 z uwzględnieniem perspektywy do roku 2024 (2015)
3. M. Crosetto et al., Subsidence monitoring using SAR interferometry: reduction of the atmospheric effects using stochastic filtering, Geophysical Research Letters, 29(9), 26-29 (2002)
4. T. Strozzi, U. Wegmiiller, L. Tosl, G. Bitelli, V. Spreckels, Land Subsidence Monitoring with Differental SAR Interferomerty, Photogrammetric Engineering & Remote Sensing, 67, 1261-1270 (2001)
5. H. Duzgun, N. Demirel, Remote Sensing of the Mine Environment, CRC Press, (2011)
6. Z. Perski, Applicability of ERS-1 and ERS-2 InSAR for land subsidence monitoring in the Silesian coal mining region, Poland, Int. Arch. Photogramm. Remote Sens., 32, 555-558 (1998)
7. E. Pilecka, D. Szwarkowski, An application of the ground laser scanning to recognise terrain surface deformation over a shallowly located underground excavation, E3S Web of Conference, Vol. 24, 1-8 (2017)
8. R. P. Gupta, Remote Sensing Geology, Springer-Verlag Berlin Heidelberg New York, (2003)
9. R. F. Hanssen, Radar Interferometry. Data Interpretation and Error Analysis, Kluwer Academic Publishers (2001).
10. A. Ferretti, F. Novali, R. Bürgmann, G. Hilley, C. Prati, InSAR Permanent Scatterer Anaysis Reveals Ups and Downs in San Francisco Bay Area, Eos, 85, No. 34, 317-324 (2004)
11. P. Prats-Iraola et al., Sentinel-1 Assessment of the Interferometric Wide-Swath Mode. 1-4. 10.1109/IGARSS.2015.7327018 (2015)
12. A. Ferretti, A. Monti-Guarnieri, C. Prati, F. Rocca, InSAR Principles: Guidelines for SAR Interferometry Processing and Interpretation, ESA Publications, TM-19 (2007)

The influence of the fault zone width on land surface vibrations after the high-energy tremor in the "Rydułtowy-Anna" hard coal mine

Elżbieta Pilecka[1,1], *Dariusz* Szwarkowski[1]

[1]Cracow University of Technology, Faculty of Civil Engineering, Krakow, Poland

Abstract. In the article, a numerical analysis of the impact of the width of the fault zone on land surface tremors on the area of the "Rydułtowy – Anna" hard coal mine was performed. The analysis covered the dynamic impact of the actual seismic wave after the high-energy tremor of 7 June 2013. Vibrations on the land surface are a measure of the mining damage risk. It is particularly the horizontal components of land vibrations that are dangerous to buildings which is reflected in the Mining Scales of Intensity (GSI) of vibrations. The run of a seismic wave in the rock mass from the hypocenter to the area's surface depends on the lithology of the area and the presence of fault zones. The rock mass network cut by faults of various widths influences the amplitude of tremor reaching the area's surface. The analysis of the impact of the width of the fault zone was done for three alternatives.

Keywords: fault zone, surface vibrations, high-energy tremor

1 Introduction

1.1 Character of high-energy mine tremors in the area of GZW

Urbanized areas of the Upper Silesian Coal Basin (*Górnośląskie Zagłębie Węglowe, GZW*) are exposed to tremors induced by underground mining operations. The tremors which are most dangerous for buildings on the area's surface are high-energy tremors with the energy limit of above $E \geq 10^5$ J for GZW. Particularly dangerous are the tremors with an energy of above $E \geq 10^7$ J. Low-energy tremors are related directly to hard coal mining and do not cause damage on the area surface. High-energy tremors have a slip mechanism in the tremor epicenter and are the result of interactions between tectonic and operational tensions [1,2,3,4,5,6,7]. The nature of the slip mechanism in the hypocenter of the high-energy tremor is similar to that in the formation of natural earthquakes. High-energy tremors cause mining damage on the area surface and have an adverse effect on the comfort of living of GZW residents. There are faults in the rock mass which cause distortions in the propagation

1 Corresponding author: epilecka@pk.edu.pl

of the seismic wave after a high-energy tremor. The article assesses the impact of the width of the fault zone in land vibrations on the area's surface. The analyses was performed after on actual signal recorded after a high-energy tremor in the hard coal mine (KWK) "Rydułtowy – Anna" on 7 June 2013 with an energy of E = 8.12e7 J. The computational model was based on a simplified actual cross-section along the Kolejowy fault within the mining area of "Rydułtowy – Anna" with a width of the fault zone of 30 m [8]. For comparison, calculations were done for other fault zone alternatives.

1.2 Threats to buildings in the area of GZW

The impact of the soil vibrations caused by mining tremors on the area's surface depends on the energy of the tremors, the epicentral distance and the structure of the rock mass through which the seismic wave passes [8]. A measure of this effect is the intensity of the vibrations which may be expressed in the degrees of intensity according to empirical scales, which has its equivalent in the maximum values of the velocities of accelerations of the vibrations. Vibrations caused by tremors induced by mining operations are recorded on the earth surface by means of seismometric sensors. These sensors measure, in three perpendicular directions, the amplitudes of acceleration, velocity or displacement of vibrations of soil particles in the function of time. The impact of the vibrations originating from mining tremors on buildings is determined through the actual runs of the vibrations at the point on the land surface which is examined or in a building. From the viewpoint of the influence of mining tremors on buildings, most important are the horizontal components x, y of the vibrations [9]. The relevant figures of the vibration parameters which were measured were used to design empirical scales of the impact of tremors induced by mining operations on building structures. In the Polish conditions, the dynamic impact of mining tremors is established using the empirical Mining Scales of Intensity (GSI) developed at the Central Mining Institute in cooperation with other institutions. For the area of GZW, the vibrations scale GSI-GZW_{KW} [10] was developed in 2007. The GSI-GZW_{KW} is an empirical scale based on the amplitude of the velocity of the horizontal land vibrations the while they continue to appear. The GSI-GZW_{KW}-2012-V was an improvement of the GSI-GZW_{KW}. The GZW_{KW}-2012-V scale takes account of the technical condition of the buildings within the area of the GZW (e.g. the timber-frame, large-block, concrete-slab, brick, traditional technologies). Currently, the GSIS-2017 scale has been designed using the experience of the application of the former GSI [11] scales. This is an empirical and measurement-based scale used to monitor and assess the impact of mining tremors induced by hard coal mining on the buildings on the surface and the perceptibility of vibrations by humans. Unfortunately, the GSI does not take into account the impact of faults on the value of vibrations on the surface after high-energy tremors.

The area of "Rydułtowy-Anna" is characterized by a thick network of multi-family housing, public buildings and industrial facilities. In "Rydułtowy-Anna", there are surface points to measure accelerations of vibrations on land surface. Statistical analyses based on the actually measured data demonstrated the directionality of damage following high-energy shocks corresponding to the tectonics of the faults [8]. The main damage to buildings occurred on the front of the propagating seismic wave. Behind the fault, the amount of damage was much smaller.

https://doi.org/10.1051/e3sconf/20183602007

2 Numerical simulation

2.1 Characteristics of the area

In this article, a fragment of the area of the "Rydułtowy-Anna" coal mine was subjected to numerical simulations in order to determine the impact of the width of the fault zone on the vibrations of land on the area surface caused by the high-energy tremor on 7 June 2013 with an energy of E = 8.12e7 J. Tectonically, the study area is made of faults and folds. It is made of carboniferous formations interrupted by numerous latitudinal and longitudinal faults. Such faults always affect mining works which influence the stability of mining roadways as well [12]. Figure 1 presents the local tectonics at "Rydutowy-Anna". The locations of the sensors (Nos. 4, 6, 7/23, 8, 15/27, 21A, 28, 29, 33) recording the tremor of 7.06.2013 were also taken into account [8].

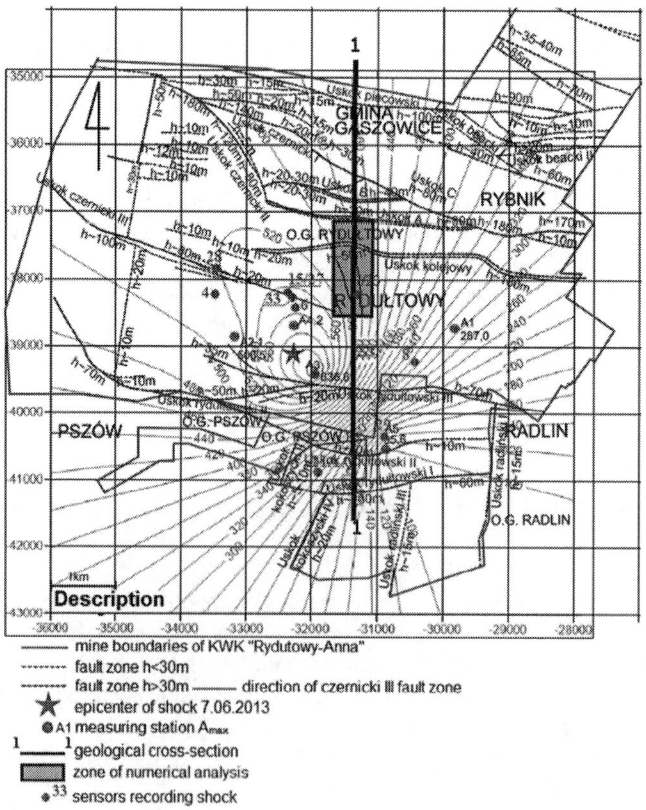

Fig. 2. The area of numerical analysis together with the location of the sensors recording the tremor of 7.06.2013 [11]

Because of the complicated tectonics of the area, the numerical simulation included the fragment which was located, in the I-I cross-section, on the marked rectangular area (Fig. 2). Only the impact of the Kolejowy fault zone of the actual width of 30 m, and two other widths of possible fault zones were taken into account: 15 m and 45 m.

2.2 Assumptions for numerical simulation

The area covered by the numerical analysis comprised the cross-section in the area of the Kolejowy fault in "Rydułtowy-Anna" coal mine, of 1100m in width and 830m in height. Because of its complicated lithological geometry, a simplified structure of the layers was adopted. The analysis covered the impact of the variable width of the Kolejowy fault on land vibrations on the area's surface before and after the fault.

The numerical model was discretized with 900000 quadrangular finite elements. The mesh size was adapted to the layer concerned, account being taken of the velocity of wave propagation in the medium and the frequency of excitations implemented in the program [13]. The material was modelled as elastoplastic and Mohr-Coulomb, and the failure criterion was employed with the linear condition of plasticity. Table 1 presents the parameters of lithological layers based on [14], as adopted in numerical simulation. Information was also included on the maximum permissible size of finite elements which was applied in the calculations.

Table 1. Parameters of the lithological layers included in the numerical simulation.

Layer:	Overburden	Shale	Sandstone	Coal 620	Zamecki sandstone	Tectonic breccia
E [MPa]	0.20	10929.00	17368.00	481.00	31980.00	0.40
ν [-]	0.30	0.26	0.22	0.32	0.22	0.30
ρ [kg/m^3]	2005.00	2324.00	2256.00	1216.00	2502.00	2150.00
ϕ [°]	38.40	30.15	44.05	7.40	52.10	21.00
c [kPa]	0.10	1800	300	120	3260	43.00
v_S [m/s]	6.19	1366.07	1776.28	387.08	2288.76	8.46
v_P [m/s]	11.59	2398.73	2964.69	752.35	3820.05	15.83
f_{max} [Hz]	50.00	50.00	50.00	50.00	50.00	50.00
h_e [m]	0.02	4.80	5.93	1.50	7.64	0.03
unit [m]	0.2	4.00	4.00	1.00	4.00	0.2

E- longitudinal displacement model, ν –Poisson's ratio, ρ –bulk density, φ – internal angle of friction, c – cohesion, v_s – velocity of the horizontal wave, v_P – velocity of the longitudinal wave, f_{max} – tremor frequency, h_e – element size.

Figure 3 presents the model implemented in the MIDAS GTS NX numerical program. Non-displaceable boundary conditions at the model base were applied. The model's own parts were set in order to determine the visco-elastic parameters of the boundary conditions, for a model with semi-seismic excitation [15]. In order to minimize the effect of wave reflection and the numerical model size, the boundary conditions of the free field on the vertical boundaries of the analyzed area [16,17] were applied. The boundary conditions which were applied take account of the Rayleigh damping.

 https://doi.org/10.1051/e3sconf/20183602007

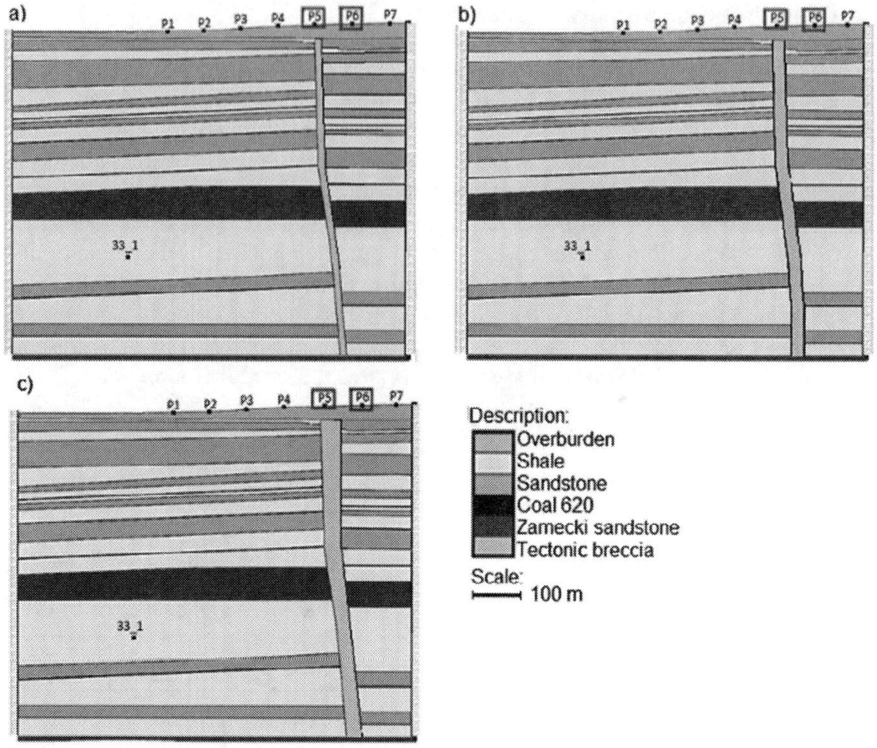

Fig. 3. Area covered by the numerical analysis. Location of points (P1 to P7) on the area surface, together with point 33_1 of the input excitation task, a) - fault zone of 15 m in width, b) – actual width of the fault zone 30 m, c) fault zone of 45 m

2.3 Semi-seismic excitation

The numerical analysis took account of the semi-seismic excitation, recorded on 7.06.2013, in the form of the course of the velocity of vibrations. Account was taken of the horizontal signal 33_X and the vertical signal 33_Z. The excitation components were implemented in the numerical model at the depth of 600 m. The load set for the model is located at a distance of 535 m from the extreme surface of the Kolejowy fault. The frequency of the recorded signal was 500Hz. The optimization of the duration of the numerical calculations necessitated the application of the discretization of the horizontal and vertical velocities of the vibrations. The time step of 0.02 s was adopted which corresponded to the signal frequency equaling 50 Hz. The duration of the excitation, introduced into the model, comprised a range of between 0.5 s and 3 s. Figure 4 presents a diagram of the velocity of vibrations of the horizontal component from the recording device, and the horizontal component introduced into the numerical model.

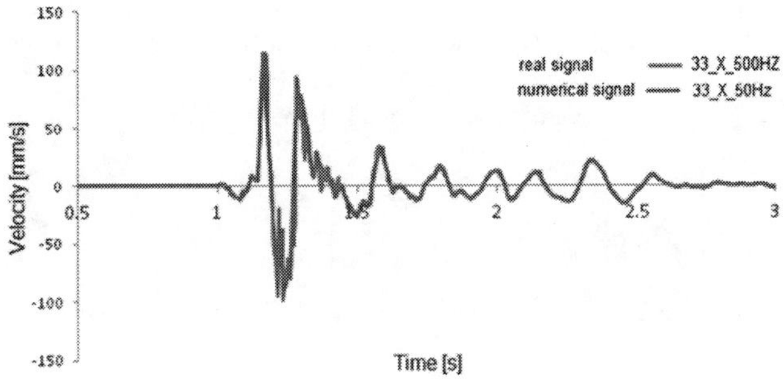

Fig. 4. Run of the horizontal component of the excitation, taking account of the frequency of sampling in the numerical program [11]

Figure 4 presents the run of the vertical component of the wave velocity recorded on the measuring sensor on 7.06.2013 and the discretized signal in the numerical program.

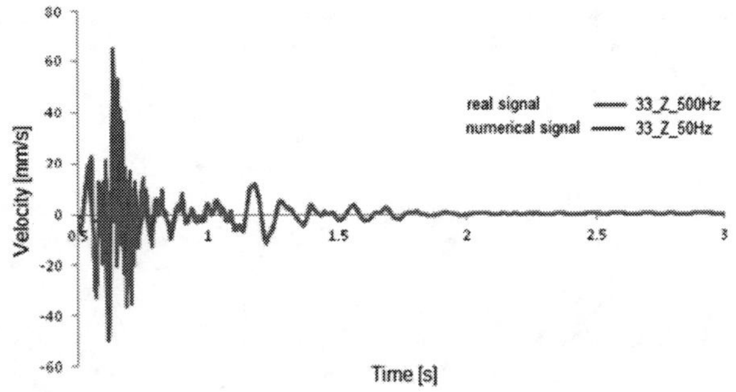

Fig. 5. Run of the horizontal component of the excitation, taking account of the frequency of sampling of the numerical program [11]

Figure 5 presents the amplitude spectra of the recorded signal. The dominating frequency was 5 Hz.

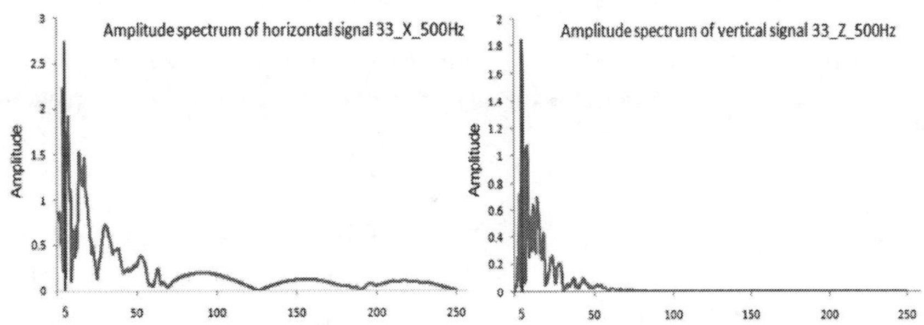

Fig. 6. Amplitude spectrum for the horizontal and vertical components of wave velocity, recorded on 7.06.2013 [11]

3 Results and final conclusions

The numerical analyses of the impact of the fault zone width on surface deformations of the areas of "Rydułtowy-Anna" demonstrated an impact of the width on the deformations generated before and after the fault. Figure 7 presents horizontal displacements obtained in point 5 before the fault. The displacement figures are close one to another. There is no clear effect of the impact of the width of the fault on the results which were obtained.

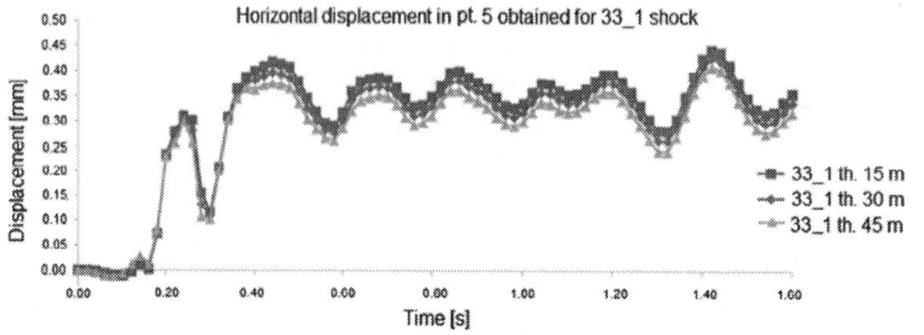

Fig. 7. Horizontal displacements at pt. 5, located on the border of the fault zone with a different thickness

Figure 8 shows the horizontal displacements obtained for pt. 6 located behind the fault zone. There is a clearly visible reduction in the horizontal displacements of vibrations for the propagating excitation in the fault zone of 45m in width. The greatest displacements were obtained for the fault zone of 15 m in width.

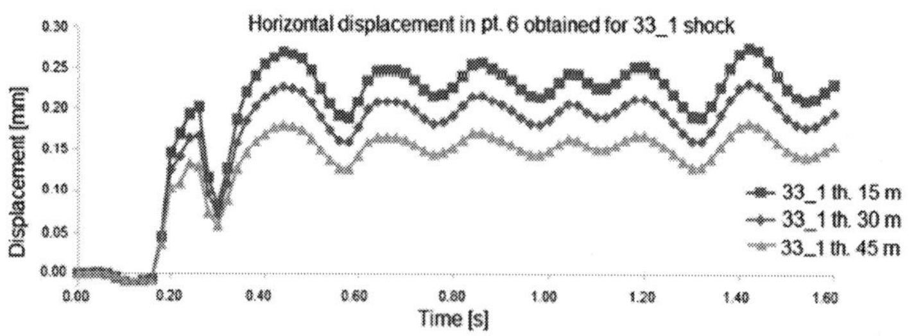

Fig. 8. Horizontal displacements in pt. 6, located after the fault zone with a different thickness

A greater width of the fault zone generates increased damping of the excitation in the form of a mining tremor. The effect of the width of the analyzed zone disappears for surface points located before the fault. The figures of total displacements are close one to another regardless of the zone width. The buildings located on mining areas after the fault zone are less exposed to negative effects of tremors induced by underground mining operations.

References

1. J. Dubiński, K. Stec, 2000: Modalność sejsmiczności górniczej w świetle badań mechanizmów ognisk wstrząsów górniczych, Publ. Inst. Geophys. Pol. Acad. S.C., M-19(281), p. 57-71;
2. S. Gibowicz, S. Lasocki, 2001: Seismicity induced by mining: ten years later, Adv. Geophysics San Diego – San Francisco – New York, Academic Press, p. 39-181;
3. A. Kijko, B. Drzężla, T. Stankiewicz, 1987: Bimodal character of extremal seismic events in Polish mines, Acta Geph. Pol., 35, p. 1157-1168;
4. H. Marcak, 1985: Geofizyczne modele rozwoju procesu niszczenia górotworu poprzedzające tąpnięcia i wstrząsy w kopalniach, Publ, Inst. Geoph. Pol, Acad, Sci. M-5, p. 149-173; in Polish;

5. E. Pilecka, 2008: Indukowanie podziemną działalnością górniczą wysokoenergetyczne wstrząsy górotworu, a lineamenty na obrazach satelitarnycz, IGSMiE PAN, Studia, Rozprawy, Monografie, in Polish;
6. Zuberek W.M. i Jochymczyk K. (red.), 2010: Geneza i charakterystyka zagrożenia sejsmicznego w Górnośląskim Zagłębiu Węglowym, Wyd. Uniwer. Śl., Katowice, in Polish; in Polish;
7. I. Skrzypczak, A. Sobkowiak, W. Kokoszka, J. Kogut, 2017: The impact of underground mining on the threat of building structures. 17th International Multidisciplinary Scientific Geoconference – SGEM, Proceedings vol 17, issue 11, Science Technologies in Geology, Exploration and Mining, pp. 623-630, Albena, Bulgaria;
8. R. Szermer-Zaucha, 2016: The impact of strong mining tremors on damage to buildings in relation to the geological and tectonic structure, doctoral thesis, Cracow University of Technology;
9. Tatara T., 2012: Odporność dynamiczna obiektów budowlanych w warunkach wstrząsów górniczych, Wydawnictwo PK, Kraków; in Polish;
10. J. Dubiński, G. Mutke, T. Tatara, L. Muszyński, A. Barański, T. Kowal, 2013: Zasady stosowania zweryfikowanej Górniczej Skali Intensywności Drgań GSIGZWKW-2012 do prognozy i oceny skutków oddziaływania wstrząsów indukowanych eksploatacją złóż węgla kamiennego w zakładach górniczych Kompanii Węglowej S.A. na obiekty budowalne i ludzi, GIG;
11. E. Pilecka, R. Szermer-Zaucha, The impact of tectonics on mining damage, monograph (to be published)
12. Małkowski P., Ostrowski Ł., Bachanek P.,2017: Modeling the small throw fault effect on the stability of a mining Broadway and its verification by in situ investigation. Energies, 10, 2082, doi: 10.3390/en10/22082.
13. Itasca Consulting Group Inc., FLAC, 2005 Fast Lagrangian Analysis of Continua. User's Guide, Minneapolis;
14. Z. Pilecki, 2011: Numerical modeling of stress field influenced by multi-seam hard coal mining in strong seismic threat conditions, The Bulletin of The Mineral and Energy Economy Research Institute of the Polish Academy of Sciences, No. 80, p. 93-102;
15. A. H. Nielsen, 2013: Towards a Complete Framework for Seismic Analysis in Abaqus, Eng. and Comput. Mech., Vol 167, EM1, p. 3-12;
16. J. Lysmer, R. L. Kuhlemeyer, 1969: Finite Dynamic model for infinite media, J ENG MECH DIV-ASCE, Vol. 95, No. EM4, p. 859-876;
17. J. Lysmer, G. Waas, 1972: Shear Waves in Plane Infinite Structures, ASCE J.Eng. Mech., 98 (EM1), p.88-105;
18. MIDAS GTS NX, Manual specifications, 2016;
19. E. Pilecka, R. Szermer-Zaucha, 2014: Analysis of the influence of Rydultowy fault on mining damages induced by tremors registered on 21 April 2011 and 7 June 2013, (Polish Mining Review, Vol. 70, No. 6, p. 60-66;
20. E. Pilecka, D. Szwarkowski, 2017: An application of the ground laser scanning to recognise terrain surface deformation over a shallowly located underground excavation, E3S Web of Conferences, Vol. 24;

The use of vibration monitoring to record the blasting works impact on buildings surrounding open-pit mines

Anna Sołtys[1,*], *Józef* Pyra[1] , and *Jan* Winzer[2]

[1]AGH University of Science and Technology, Faculty of Mining and Geoengineering, Krakow, Poland

[2]Exploconsult Sp. z o.o., Krakow, Poland

Abstract. Environmental protection law and geological and mining law require the mineral mining plant to protect its surroundings from the effects of mining operations. This also applies to the negative impact of vibrations induced by blasting works on people and construction facilities. Effective protection is only possible if the level of this impact is known, therefore it is necessary to record it. The thesis formulated in this way has been and continues to be the guiding principle of the research works carried out in the AGH Laboratory of Blasting Work and Environmental Protection. As a result of these works are procedures for conducting preventive activities by open-pit mines in order to minimize the impact of blasting on facilities in the surrounding area. An important element of this activity is the monitoring of vibrations in constructions, which is a source of knowledge for excavation supervisors and engineers performing blasting works, thus contributing to raising the awareness of the responsible operation of the mining plant. Developed in the Laboratory of the Mine's Vibration Monitoring Station (KSMD), after several modernizations, it became a fully automated system for monitoring and recording the impact of blasting works on the surrounding environment. Currently, there are 30 measuring devices in 10 open-pit mines, and additional 8 devices are used to provide periodic measurement and recording services for the mines concerned.

Keywords: open-pit mining, millisecond blasting, vibration monitoring

1 Introduction

Initiating the explosives in long holes is a commonly used method of rock mining, and that is why the related problems are so important for open-pit mining. Only 20 to 30 percent of the detonation energy of explosives is used for mining. The remaining part is lost and it causes scattering of flyrocks, airblast, and seismic waves. Especially the latter ones

* Corresponding author: soltys@agh.edu.pl

cause a lot of problems because the vibrations of the substrate can affect various building structures in the vicinity of mines. Both mining efficiency and economic aspects lead to the use of as many explosive charges as possible in a single batch, which may, however, cause the increasing problem of harmful impacts.

Long-term research work (also financed by mines) carried out at AGH's Department of Open-pit Mining and activities aimed at disseminating the idea of preventive action in this field in mining plants [1, 2, 3], allowed on the one hand to develop specific research procedures, and on the other hand to achieve a certain level of awareness among mining supervisors, which results in ongoing consultations, conducting control measurements or even full monitoring of the blasting impact on the environment. This decisively increases the mental comfort of both supervisors and investigators and analysts. Such actions are in most cases socially acceptable.

Generally, the preventive activity of open-pit mines covers two groups of the issue [4]: basic research and recording the level of vibrations in the environment.

The aim of basic research is to determine the conditions for safe execution of blasting works taking into account local geological and mining conditions, as well as the type, quality and technical condition of the surrounding buildings. The final effect of the basic research is to determine the dependencies allowing for the calculation of the allowable mass of explosives for the expected range of excavation within a defined time interval, together with the description of the technique and technology for performing blasting works.

Recording of the blasting work's impact is aimed at checking the validity of the relations specified in the basic research, i.e. the compliance of the intensity level of vibrations induced in the environment with the predicted values together with the assessment of the impact of the recorded vibrations on the buildings. These activities may be performed, as cyclical control measurements or in a wider scope, as vibration monitoring in designated buildings in the vicinity of mining operations.

It results that prevention activities in this respect should include:
- an inventory of the technical condition of the facilities in the surroundings,
- recognition of the directions of vibration propagation and their intensity in the vicinity of the mine excavation, on the basis of which the range of harmful vibration impact and allowable mass of explosives are determined,
- conducting cyclical control measurements,
- vibration monitoring in protected objects in special cases.

The last two points mentioned above are repeatedly cited by experts as recommended for implementation and, in the last period of time, also required by the concession granting authorities as a condition for obtaining it. These are activities resulting from the Environmental Protection Law and the Geological and Mining Law, applicable to enterprises that may have a significant impact on the environment. These include the majority of mines extracting minerals using explosives.

2 Monitoring of the effects of vibrations on the environment

Monitoring of vibrations induced by blasting works was the subject of research and projects carried out since 1996 in the Department of Open-pit Mining at AGH [2] and carried out in cooperation with industry. The result of works was the implementation of the measurement, analytical and archiving apparatus into the mining practice. When designing the first measurement system, it was assumed that the measurement of vibrations must be supplemented by an assessment of impact using the Polish standard. This is why the system was equipped with a software, which made it possible to use correction filters complying with the PN-85/B-02170 [5] standard, which made it possible to assess the impact immediately after the measurement. Practically this procedure of system operation is still being used, obviously taking into account new analytical and graphical abilities. The cooperation of the AGH's Department with Exploconsult Sp. z o. o. and later also with A-STER s. c. allowed for rapid development of the system to its full automation and adjustment of the impact assessment to the requirements of PN-B-02170:2016-12 [6] standard.

In open-pit mining, the following measuring systems have been used to monitor vibrations induced by blasting works [2]:
- Small Vibration Monitoring Station - Explo 504,
- Vibration Monitoring Mine in 3 versions:
- KSMD with radio communication,
- KSMD GSM with cellular communication,
- KSMD APN with the Internet connection (fig. 1).

The main idea of the changes was to make the system more flexible by:
- direct contact between the measuring station and the central server,
- limitation of the service by the operator,
- the use of the wireless Internet connectivity,
- data is collected on the server and can be accessed at any time without the need to establish a connection to the measuring station.
- remote service maintenance and changes in the software of the measurement device.

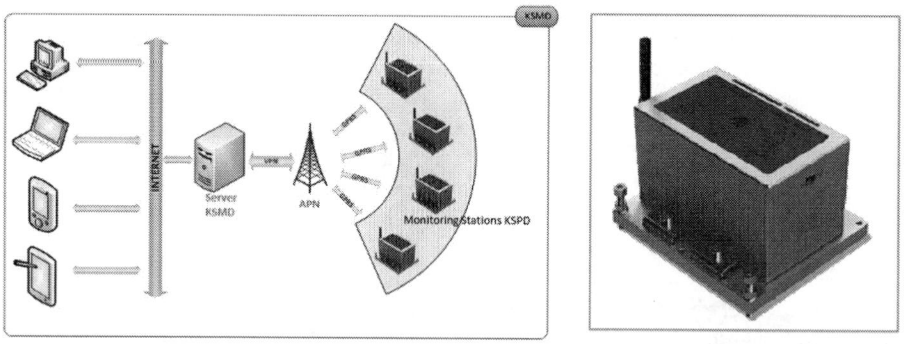

Fig. 1. Mine Station of Vibrations Monitoring - 2012 model.

The basic link of the new system is the KSPD Mine Vibration Measurement Station (fig. 1), equipped with a 3-axis vibration velocity meter and a number of processing and measurement data collection systems. An integral part of the station is a GSM/GPRS modem that automatically sends the collected measurements to a KSMD server located at AGH, in Krakow.

The web service is used to manage the stations and measurements made by them. It can be used to view measurements, generate reports, manage devices, buildings, and system users. An access to the service is protected by a username and password.

The most important advantage of the modernised system is collecting measurement data on the server - KSPD measuring station, after saving the event in memory, automatically establishes contact and sends the data to the server. In case of communication problems, data transmission shall be repeated until the successful transmission. This is very important because KSPD does not store the data but the KSMD server does. As a result, the system operator cooperates only with the server in terms of analysing or using the database.

Advantages of the new KSMD system:
- continuous vibration measurement in the constructions,
- unlimited memory,
- automatic data transmission to the server,
- constant connection with the server,

- automation of the station starting process,
- the possibility of contact from any point on the Earth,
- archiving of complete vibration waveforms,
- archiving of reports and impact assessments,
- visualisation of vibration waveforms in time and results on the SWD scale,
- reporting after each event.

Currently, there are 42 measuring points in 13 open-pit mines:
- KSMD APN model - 10 mines and 31 Mining Vibration Measurement Stations (KSPD),
- KSMD GSM model - 2 mines and 11 measurement points.

In addition, AGH Laboratory of Blasting Works and Environment Protection has 2 KSPD stations and Exploconsult company - 7 stations. These stations are periodically rented to perform monitoring in open-pit mines as well as to control the impact of engineering works using explosives (e.g. macro-levelling during the construction of roads, tunneling and demolition works).

3 Analysis of recording from blasting works

The operation of the modernised KSMD system started in 2012, therefore the five-year period of operation of the system can be summarised.

In total, all KSMD ANP measurement stations performed 11407 measurements of vibrations induced by blasting works carried out in open-pit mines during last 5 years (2013 to 2017).

Tables 1 and 2 show the number of measurements made by the measurement stations in individual mines until the 31st of December 2017.

Table 1. Vibration measurements made at individual mines - stations owned by mines

Mine / raw material	Starting date	Number of measuring points	Number of measuring points (facilities)	Number of measurements
Gypsum mine 1	05.07.2012 r.	4	8	1491
Limestone mine 1	04.07.2012 r.	3	3	1791
Limestone mine 2	29.10.2012 r.	4	5	1387
Limestone mine 3	25.05.2013 r.	4	5	452
Gypsum mine 2	09.04.2013 r.	4	5	1357
dolomite mine 1	15.04.2013 r.	2	5	1521
Limestone mine 4	14.03.2014 r.	2	3	695
Limestone mine 5	19.07.2016 r.	1	1	136
dolomite mine 2	23.11.2016 r.	2	2	833
shale mine	24.05.2015 r.	2	2	101
In total				9764

In all cases KSPD measuring stations are fixed to building foundations at ground level, i.e. the vibration measurement concerns a building and not the ground in its surroundings. This is important information, as the standards of other countries set specific but different requirements in this respect. For example, in Poland, the Mining Rocks of Intensity require the measurement of ground vibrations.

Table 2. Vibration measurements made at individual mines - measurements carried out as a part of the service

Mine / raw material	Operation period	Number of measuring points	Number of measuring points (facilities)	Number of measuremen ts
dolomite mine 3	since 04.08.2015 -	1	7	160
Limestone mine 6	since 23.08.2016	4	4	1233
granite mine	15.05.2015 to 31.07.2015	1	1	24
dolomite mine 4	02.09.15 to 10.02.16	1	1	52
basalt mine	25.05.16 to 26.06.17	1	1	44
ampibolite mine	16.08.16 to 23.03.17	2	2	30
Limestone mine 7	03.11.16 to 22.02.17	1	1	86
dolomite mine 5	09.06.17 to 30.06.17	1	1	14
In total				1643

For example, the results of measurements obtained by the station - limestone mine 2 / station 1 were analysed in details. Table 3 presents the part of recording results for the 1 quarter of 2013. This part was selected for recording, on January the 17th, the events with the highest intensity during 5 years of station operation.

Table 3. Results of vibration recording in the first quarter of 2013 - limestone mine - station No. 1

Date	Time	Vibration velocity, mm/s			Frequency, Hz		
		u_z	u_x	u_y	f_z	f_x	f_y
2013-01-09	13:25:22	0,71	1,15	0,98	11,0	9,6	11,6
2013-01-17	**13:20:37**	**1,52**	**2,64**	**5,20**	**12,6**	**11,4**	**11,0**
2013-01-17	13:23:05	1,24	2,64	2,20	13,0	12,0	11,6
2013-02-14	13:21:44	2,08	3,43	2,44	17,0	16,7	11,3

The figure 2 presents the visualization of recordings on the SWD-I scale (horizontal components x and y), and the percentage distribution of maximal velocities within the assumed value ranges.

The visualization of the measurement results is shown on two graphs, as the maximum value of the vibration velocity (without frequency) does not give full information about the impact, due to the shape of the SWD limit lines. At the same time, it is important to be

 https://doi.org/10.1051/e3sconf/20183602008

aware that the measurement results, in the form of maximum values of frequency correlated with vibration velocities, applied on the SWD scale do not constitute an impact assessment, but only illustrate the scale of the problem and allow for the preliminary conclusions to be drawn as to the level of recorded vibrations over a longer period of time.

When analysing the data presented in Figure 2, it should be noted that almost 85% of the vibrations induced by the blasting works did not exceed a maximum value of 1.33 mm/s, which means that they can be classified as negligible in the process of impact assessment for a protected object [6].

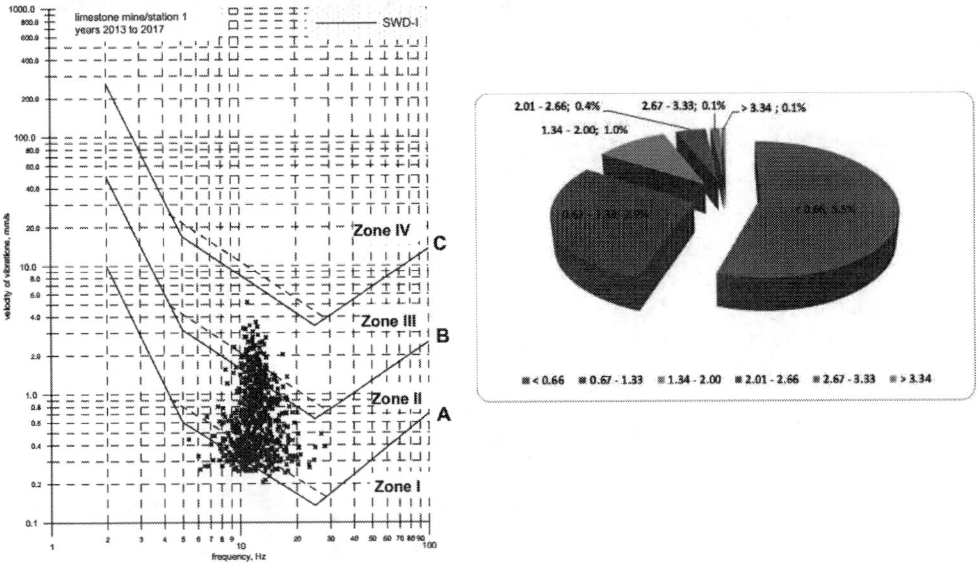

Fig. 2. Visualization of vibration recording results for the years 2013 to 2017.

As already mentioned, on 17 January 2013, the event with the highest intensity during the five-year period was recorded. For example, the structure of these vibrations are shown in Figures 3. As shown in Figure 3, the 12.59 Hz frequency is predominant in the structure of vibration and the horizontal component y is predominant in the impact assessment, as shown in Figure 4.

The assessment of induced vibrations impact, in accordance with the procedure in the standard [5] (fig. 4), has shown that these vibrations should be classified as zone III of the SWD-I scale. It should be noted here that it is a single event recorded over a period of 5 years, and only 0.6 % of the recording (4 events out of 692) exceeded the value of 3.33 mm/s.

 https://doi.org/10.1051/e3sconf/20183602008

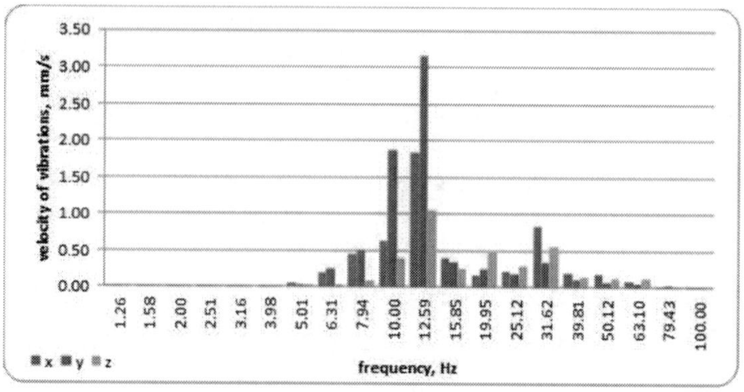

Fig. 3. The structure of vibrations registered on 17.01.2013 - limestone mine / station 1.

The number of performed measurements and the scope of measured values are the information of the essential significance during every compensation proceeding. The fact that most events are recorded by vibration measurements, each of which can be used as a basis for assessing the impact on the building, and most importantly, only a few (of the highest intensity) of the hundreds of documented events were perceived vibrations, is very important for the opinion formers, and consequently also for the courts.

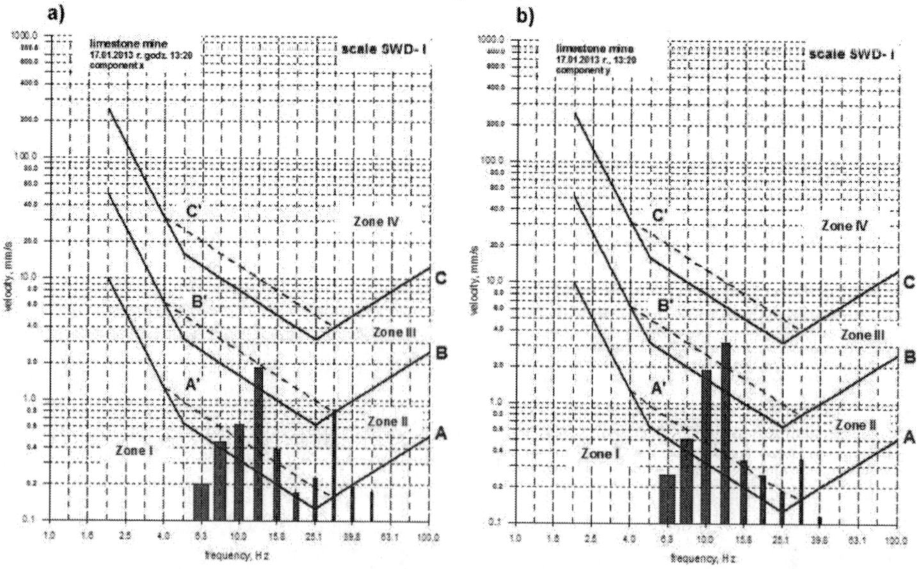

Fig. 4. Vibration impact assessment - limestone mine/station 1 - event dated of the 17th of January 2013: a) component x, b) component y

On the basis of observations made during the monitoring, despite the occurrence of recorded vibrations which were qualified to the third zone of the SWD scale, no new damages or deepening of already existing cracks in the building structure has been observed. It has to be emphasized that this was the only event qualified to the third zone, out of 692 recorded within 5 years window. It can, therefore, be concluded that the occurrence of individual events qualifying to the third zone of the SWD scale does not contribute to the occurrence of new damages in the protected site.

4 Summary

Monitoring of vibrations induced by blasting works on constructions, in the vicinity of the excavation, is an important element of the prophylactic activity of open-pit mines in the scope of minimizing the impact of exploitation works on the environment.

Documenting the impact of blasting works on the environment is a solution that allows to:

- collect up-to-date information for surveillance, concerning the intensity of vibrations induced by blasting works,
- monitor the impact of vibrations on building structures,
- create a database, which can always be used to provide evidence of damages,
- make current corrections under conditions limiting the performance of blasting works.

The KSMD system used in mining plants, which was built with the financial participation of open-pit mines, has been gradually modernized, thanks to which it is now a modern, automated system which uses new technologies in the field of measurement, analysis and wireless communication. The system has been created as a tool for supervising personnel, enabling the current control of the blasting works impact on buildings in the surrounding area.

After the modernization, KSMD enables access to measuring devices from any point on the ground, control of measuring equipment, an immediate preview of the recorded event and rapid impact assessment.

In vibration monitoring nothing is accidental. The permanent presence of measuring equipment in the protected object allows for recording also events not related to blasting works. Sometimes shocks caused by underground mines are recorded, as well as vibrations induced by blasting works in neighbouring open-pit mines.

The operation of vibration monitoring systems provides thousands of evidence that controlled blasting works in open-pit mines can be harmless to buildings in the surrounding environment.

References

1. E. Maciąg, J. Winzer, R. Biessikirski, Metodyka postępowania w ochronie otoczenia w przypadku robót strzałowych. Bezpieczeństwo Pracy i Ochrona Środowiska w Górnictwie 9(157)/I/2007, pp. 56 – 60 (2007)
2. J. Winzer, Dokumentowanie oddziaływania robót strzałowych na otoczenie kopalń odkrywkowych. Górnictwo Odkrywkowe, R. 48, nr 3-4/2006, pp. 159-167 (2006)
3. J. Winzer, A. Sołtys, J. Pyra, Oddziaływanie na otoczenie robót z użyciem materiałów wybuchowych. (Wydawnictwa AGH, Kraków, 2016)
4. J. Pyra, A. Sołtys, J. Winzer, Skomputeryzowane systemy do dokumentowania oddziaływania robót strzałowych na otoczenie kopalń odkrywkowych. Bezpieczeństwo Robót Strzałowych, GIG, pp. 79 – 88 (2012)
5. PN-B-02170:1985 - Ocena szkodliwości drgań przekazywanych przez podłoże na budynki.
6. PN-B-02170:2016-12 – Ocena szkodliwości drgań przekazywanych przez podłoże na budynki.

Assessment of deformations in mining areas using the Riegl VZ-400 terrestrial laser scanner

Dariusz Szwarkowski[1*], *Magdalena* Moskal[1]

[1]Cracow University of Technology, Faculty of Civil Engineering, Krakow, Poland

Abstract. The article discusses the use of terrestrial laser scanning to assess deformations in mining areas. Using the terrestrial laser scanning Riegl VZ-400, control measurements within the historical location of the underground coal mine in Zabrze were made. Two laser scanning measurements were taken over the course of one year. The research made it possible to determine changes in surface deformation on the shallowly located mining excavations. Differences in the terrain may be due to subsidence associated with the influence of underground mining and pose a threat to the adjacent road infrastructure and structures.

Keywords: terrestrial laser scanning, ground deformation, USBC

1 Introduction

The urbanized areas of the Upper Silesian Coal Basin (USCB) are characterized by deformations resulting from underground coal mining. In order to verify its impact on the existing road infrastructure and buildings, the endangered areas should be subjected to continuous monitoring. Changes in the surface area caused by underground mining excavations may take place even after a long period of time after the mining ended. The most common form of assessment of changes occurring in the rock mass is the deformation of the surface layers of the ground and the deformations of the terrain.

There are several ways to determine the measure of deformation of subsurface layers, among others with the use of benchmarks, satellite, aerial (LIDAR) and ground remote sensing [1,2].

The paper described the RIEGL VZ-400 terrestrial laser scanner characterized by ease of use, independence of measurements over time and sufficient accuracy of results. Currently, terrestrial laser scanning is commonly used in applications in the reclamation of post-mining areas [3,4]. It is the most frequently used form of monitoring mining deformations.

The paper discusses the important topic of land deformation in post-mining, strongly urbanized areas. The subject of considerations is the area over a series of underground mining excavations in Upper Silesia. In order to determine the surface deformation there, two measurements were made using a terrestrial laser scanner at an interval of ca. one year .

* Corresponding author: dszwarkowski@pk.edu.pl

In order to identify surface deformation changes, both surface images were compared. The measurements are aimed at determining the suitability of terrestrial laser scanning to recognize discontinuous surface deformations.

2 Characteristics of the Riegl VZ-400 terrestrial laser scanner

2.1 Scanning mechanism

The Riegl VZ-400 terrestrial laser scanner emits a narrow beam of infrared light during measurement. The device is equipped with a linear scanning mechanism, based on a multi-angle mirror rotating at high speed, making it possible to obtain unidirectional parallel scanning lines. The polygonal rotating mirror reflects the laser beams in different directions, creating scan lines. Along with the linear scan movement, the optical head rotates while performing a frame scan of 360 degrees. The laser pulse is collimated by the optics forming the beam and then reflected by the facet of the polygon mirror to finally pass through the exit window of the beam. Along the way, the emitted laser pulse can hit one or several targets, inducing first or another reflected pulses. These reflected "echo" optical signals run back into the device, pass through the glass window, are reflected to the receiver optics through the mirror and focused on the receiver. The receiver converts optical signals into electrical ones, which are amplified and subsequently sampled and digitized in the process of converting analogue signals into digital ones. Digitized signals are transmitted to the controller to process the online waveform [5].

2.2 Generating data in the receiver

Figure 1 shows the output signal of the emitter. The first pulse on the left comes from the laser transmitter. Encountering an obstacle in the form of eg a tree, it is reflected, which results in a partial "dampening" of the amplitude of the output pulse, similarly in the case of the extreme signal on the right of Fig. 2.1, illustrating the impulse reflected from the roof.

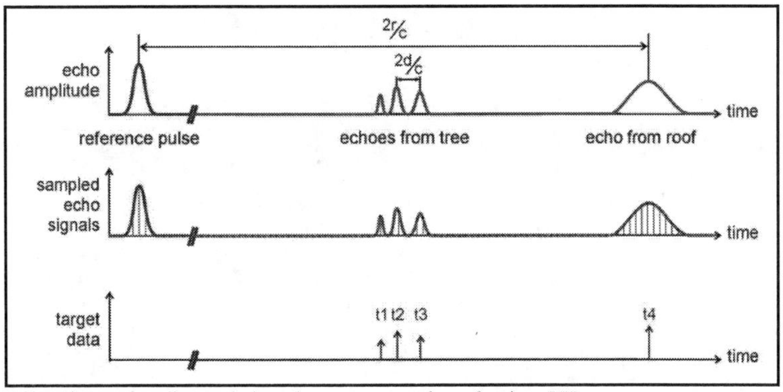

Fig. 1. Collection of reflected data by a measuring device [5].

The vertical lines in Fig. 1 correspond to the time at which the sample was taken. The scanner software analyzes the waveform. On this basis, calculations of the time positions of the reflected signals are performed, in this case as t1, .., t4, for subsequent objects 1, ..., 4. Multiplying the obtained time values by the known speed of light, the required range of the object is obtained for the four targets. The results regarding the range of objects (beam angle, date of measurement) are stored in the set of output data.

2.3 Measurement errors in scanning

The basic objective of terrestrial laser scanner research is to obtain an accurate three-dimensional image of the measured area. An important role in distorting the output image is played by the atmospheric conditions that can significantly affect the result of scans. Measurements with a terrestrial laser scanner during precipitation should be avoided beause the scan is characterized by a huge amount of interference in the course of the point cloud which also comprises, apart from the scanned objects and terrain, very small droplets of atmospheric precipitation. A big difficulty durning data processing from the obtained point cloud are areas characterized by a large amount of vegetation and tree stands, without characteristic permanent objects. The creation of scans on the basis of a point cloud created as a record of the history of changes in the position of individual leaves, boughs of trees and shrubs causes significant simplifications in the course of the shape of the terrain surface. A disadvantage of a laser measurement is a huge amount of data obtained from individual measurements. This involves the need to transform the scans and select the area which is most desirable from the viewpoint of its use. A person who makes a terrain measurement with the use of a terrestrial laser scanner should take into account the factors which are described above and which have a significant effect on the accuracy of the surface area obtained.

3 Research area

The analyzed area is located in southern Poland, in the area of the Upper Silesian Coal Basin (USCB). It is characterized by the extraction of coal, zinc, lead, iron and silver deposits since early Middle Ages [6, 7, 8]. Due to the shallow extraction of coal, this area is particularly exposed to discontinuous deformations.

Fig. 2. Location of the research area with marked mining corridors.

The measurement with a terrestrial laser scanner was carried out in Zabrze, on plot No. 4962/42, at the intersection of Końcowa and Wolności Streets (Fig. 2). A city park with numerous trees and shrubs is located there. Below the discussed area, there is a historical string of underground mining excavation "No. 8 Pochhamer cut thru". It is marked white in Fig. 1. At a depth of about 32 m, there are two parallel corridors located in Carboniferous seams. On the southern side of the excavation, a breakthrough was made in coal seams, connecting the two corridors. The described system of excavations is conducive to the formation of zones of weakening of rock mass and surface deformations of the area.

4 Analysis and results

The measurements carried out on 7 February 2018 are a continuation of the monitoring of the surface area conducted in previous years (i.e. 20 April 2015, 23 February 2017 and 1 June 2017) [8, 9]. The current measurement was performed with the Riegl VZ-400 terrestrial laser scanner at seven measurement stands. As indicated in Fig. 1, the measuring points (S1-S7) are located in the immediate vicinity of an underground mining excavation. Two meaningful tests were used to analyse the ground deformation, carried out at one-year time interval (23 February 2017 and 7 February 2018). Measurements carried out at the same time of the year enable the elimination of factors distorting the measurement and also the subsequent interpretation of the results (among others, differences in flora). Figure 3 shows the initial image obtained after the registration of scans from individual measurement stands dated 7 February 2018.

Fig. 3. Example of an image formed as a result of connecting scans from measurement stands in the research area (7.02.2018)

The analysis included a comparison of the land area obtained from both measurements in terms of its deformation and the location of their occurrence in relation to the location of underground mining excavations. Based on measurements made on 23 February 2017, a surface model was developed for Fig. 3. The terrain surface can be considered relatively flat, within the limits of 307.50 - 309.688 meters above sea level [9]. The largest deformation of the area can be seen in the area of the upper corridor of the mining excavation (Fig. 4). The zone that has settled is about 15 m long and 5 m wide.

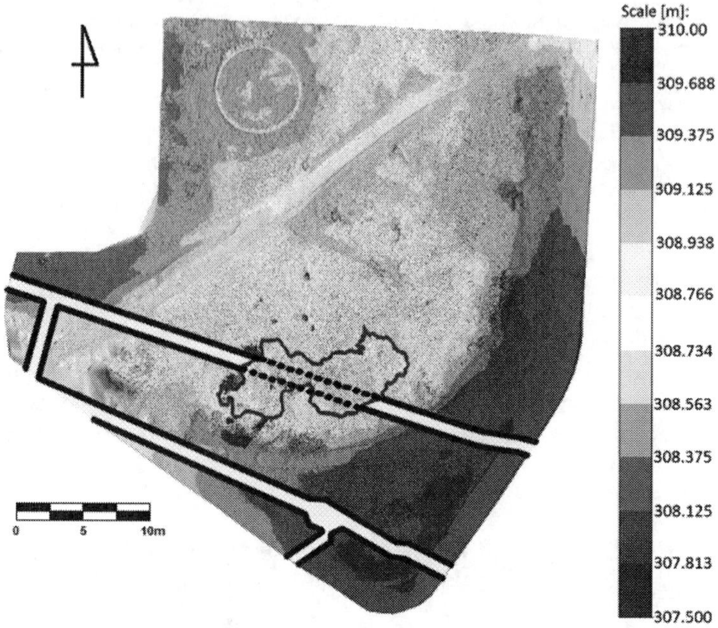

Fig. 4. Terrain surface model based on laser measurements of 23 February 2017

Another measurement of the analysed area was performed on 7 February 2018. The obtained terrain surface model is shown in Fig. 5.

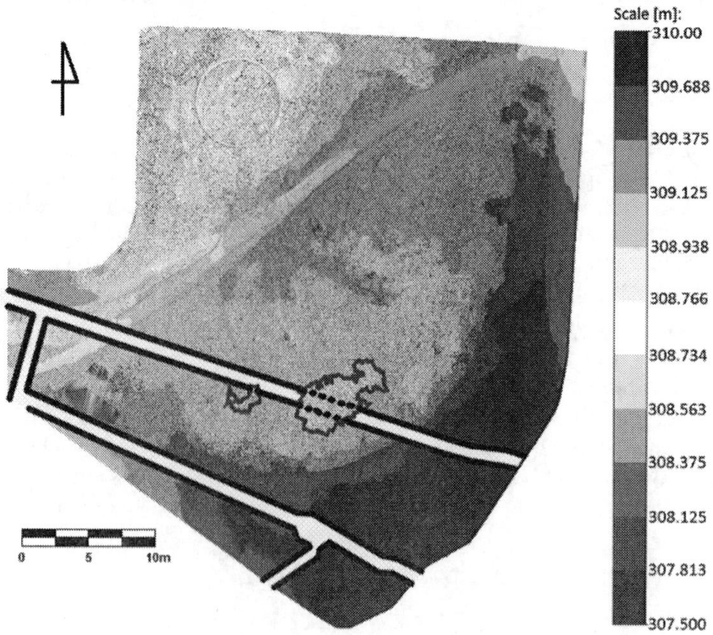

Fig. 5. Terrain surface model based on laser measurements of 7 February 2018.

The results are similar to those obtained in 2017. The visible local settlement of the area is located exactly in the same place. The area covered by the anomaly is smaller compared to the one obtained in 2017.

Based on the measurements carried out on 23 February 2017 and 7 February 2018, a differential model of the analyzed area was made (Fig. 6).

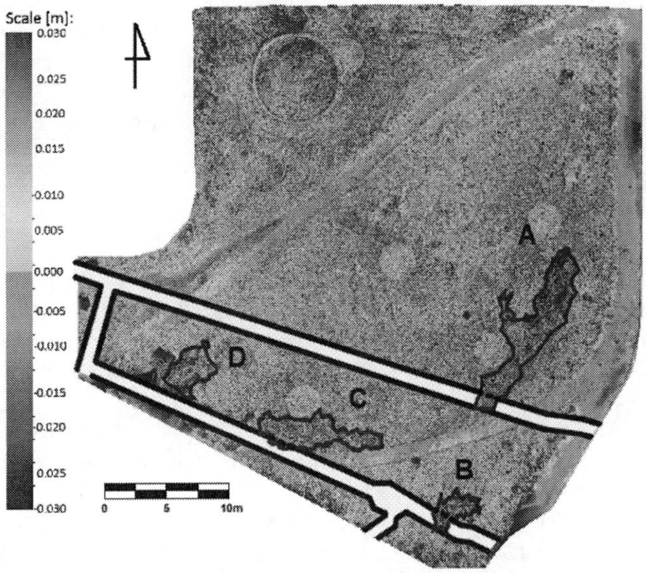

Fig. 6. Differential model of terrain surface based on laser measurements from February 23, 2017 and February 7, 2018

The calculated differences in coordinate values of the terrain surface are up to 3cm. In Fig. 6, we can see four zones [A, B, C, D] depicting possible terrain deformations related to the existence of underground mining excavations.

5 Summary and conclusions

The monitoring of analysed area by the RIEGL VZ-400 terrestrial laser scanner made it possible to determine surface deformations caused by underground mining from shallow historical mining excavations. Although hard coal is not extracted the analysed area anymore, slight deformations of the area still occur in the region of mining excavations. The monitoring carried out at a one-year interval enabled a reliable estimation of surface deformations of the area. They did not exceed 3 cm locally. Most likely, the resulting deformations are caused by subsidence of geological strata affected by the mining activity. The use of a terrestrial laser scanner made it possible to accurately reproduce the changes taking place in the spatial relief of the terrain. Due to the fast measurement time and the non-invasive monitoring, the terrestrial laser scanner measurement forms an important advantage of surface discontinuity testing. Its use is particularly beneficial in highly urbanized areas exposed to mining activities. Terrestrial laser scanner makes it possible to verify ground deformation and deformation of buildings which increase over time. This enables an assessment of the risk to the existing infrastructure and residents.

References

1. Goszcz, Powstawanie zapadlisk i innych deformacji nieciągłych powierzchni na obszarach płytkiej eksploatacji górniczej. Mat. Konf. Szkoła Eksploatacji Podziemnej '96. CPPGSMiE PAN, Szczyrk, pp. 119-137 (1996)
2. M. Chudek, Geomechanika z podstawami ochrony środowiska górniczego i powierzchni terenu. Wyd. Politechniki Śl., Gliwice (2002)
3. Miłkowski, Z. Pilecki, K. Kłosek, M. Tondera, Autostrada A1 zaprojektowana na „dziurawym" podłożu, Cz. 1. Magazyn Autostrady: budownictwo drogowo-mostowe 3/2010, pp. 104–112 (2010)
4. Z. Pilecki, A. Kotyrba, Problematyka rozpoznania deformacji nieciągłych dla potrzeb projektowania konstrukcji drogowych na terenie płytkiej eksploatacji rud metali. Prace Naukowe GIG III/2007, pp. 379-392 (2007)
5. Riegl LMS, Oprogramowanie Systemowe i Przetwarzania Danych RiSCAN PRO dla skanerów laserowych 3D firmy RIEGL LMS. Austria (2009)
6. Z. Pilecki, A. Kotyrba, Problematyka rozpoznania deformacji nieciągłych dla potrzeb projektowania konstrukcji drogowych na terenie płytkiej eksploatacji rud metali. Prace Naukowe GIG III/2007, pp. 379-392 (2007)
7. Z. Pilecki, E. Pilecka, J. Pszonka, J. Stanisz, J. Morman, Zagrożenia deformacjami nieciągłymi na obszarze górniczym Jaworzno III, IGSMiE PAN, Kraków (2012)
8. Z. Pilecki, E. Popiołek, Wpływ eksploatacji rud na zagrożenie powierzchni deformacjami nieciągłymi i jego badanie za pomocą metod geofizycznych, Studia, Rozprawy, Monografie 84, IGSMiE PAN, Kraków (2000)
9. E. Pilecka, D. Szwarkowski, An application of the ground laser scanning to recognise terrain surface deformation over a shallowly located underground excavation, E3S Web of Conferences 24 (2017)

E3S Web of Conferences **36**, 02010 (2018) https://doi.org/10.1051/e3sconf/20183602010
BIG 2018

Measurement methods of building structures deflections

Magdalena Wróblewska[1,*]

[1]Silesian University of Technology, Faculty of Civil Engineering, Department of Geotechnics and Roads, Gliwice, Poland

Abstract. Underground mining exploitation is leading to the occurrence of deformations manifested by, in particular, sloping terrain. The structures situated on the deforming subsoil are subject to uneven subsidence which is leading in consequence to their deflection. Before a building rectification process takes place by, e.g. uneven raising, the structure's deflection direction and value is determined so that the structure is restored to its vertical position as a result of the undertaken remedial measures. Deflection can be determined by applying classical as well as modern measurement techniques. The article presents examples of measurement methods used considering the measured elements of building structures' constructions and field measurements. Moreover, for a given example of a mining area, the existing deflections of buildings were compared with mining terrain sloping.

Keywords: deflected buildings, measurement methods, surface deformation

1 Introduction

Vertically deflected structures pose a serious problem in the social and technical context for building engineers at the stage of building rectification. In addition, for mining areas, where deflection values are substantial, this phenomenon is also becoming a part of financial burdens for a mining enterprise because relevant compensation has to be paid or rectification costs covered.

A deflection is a damage affecting the use of a structure. Investigations aimed at identifying users' response to building deflection were carried out, among others, under the works [1]. It was observed in the undertaken investigations that the use of a structure is becoming more difficult with the deflection value starting already from 10‰. Vertically deflected structures are additionally exposed to the change in the system of internal forces, which - for significant deflection values - affects structural safety. The permitted deflections of buildings on a mining area were analysed thoroughly in [2] due to the conditions of ultimate load-carrying limit states [2].

* Corresponding author: magdalena.wroblewska@polsl.pl

Straightening by uneven raising is a commonly applied method [3]. In such case, it is necessary to determine the adequate deflection value, considered as an input value, so as to have a structure brought to the vertical position as a result of rectification. Appropriate verticality measurement methods are employed before and after straightening, as part of a post-execution measurement, to determine the deflection value and direction. Admissible deviations in structure execution have to be known to interpret the data obtained correctly. A difference in the depth of foundations, causing building deflection of up to 3 mm/m, is permitted already at the stage of foundation design [4]. If the tolerances permitted for construction of certain structural elements determined based on standards and technical guidelines are exceeded, the accuracy of building structure execution is seriously questioned [5].

A deflection of building structures, considered as a mining damage, is reflected in the loss of the structures' value. By juxtaposing the methods applied to determine a relationship between the building deflection magnitude and the size of the damage [6], a new formula was started to be applied, based on the assumption that with each next per mille of building deflection, its incremental loss value is rising by a certain constant value.

Mining impacts on the deflected buildings, in form of land curvatures and horizontal deflections, cause the deformation of particular structural elements, as a result of which the deflection of walls is not always identical as the sloping of floors. The additional strains, formed this way, may have a negative impact on the structure's technical condition and safety [7]. Due to such differentiated values of deflection for particular structural elements of the building, a specific value and direction of structure deflection have to be determined, especially for those structures undergoing a straightening process [8].

2 Measurement methods of building structures displacements

Considering the measurement methods of building structures displacements, methods can be distinguished consisting of altimetric survey or topographical survey. In order to obtain absolute values, a measurement made has to be referred to fixed (main) points situated outside the impacts of exploitation. However, measurement often consists of determination of absolute values as there are no fixed points in the adjacency of the building. Such situations take place mainly on areas with a large number of parcels used, where the reach of the impact covers extensive areas. The number and manner of arranging controlled points on the studied structure should ensure the possibility of obtaining a full picture of the structure deformation course [9]. Intervals between measurements, i.e. measurements appropriately planned in time, allow to observe a condition of the building in the given phases of the depression basin course.

Data about deflections is collected on the basis of field measurements. The appropriate method is chosen, among others, depending on the expected accuracy, availability of equipment, software and time, availability of measurement points and visibility. The methods encompass classical measurements where distance, angles and height difference between points is measured directly, and modern methods using satellite techniques, photogrammetry and laser scanning. As technology is advancing, the automation of geodesic measurements has also been seen, including telemetric methods, the example of which can be a laser sensor of building vibrations and deflections [10].

The article presents building measurement methods for determining the deflection of structural elements, in particular walls, floors and terrain sloping measurement.

2.1 Registration of wall deflections

2.1.1 Levelling measurements

The deflections of buildings in field should be determined by means of geodesic levelling techniques by measuring the depression of the benchmarks situated on the building base level [1] (Fig. 1). Precise or technical direct levelling can be distinguished for the commonly used methods. It is very often seen in field that only single controlled points are stabilised, which provides an incomplete view of building displacement. It is therefore recommended to establish at least four controlled points in the corners of the structure [11]. In practice, this only takes place for large structures such as industrial buildings or churches.

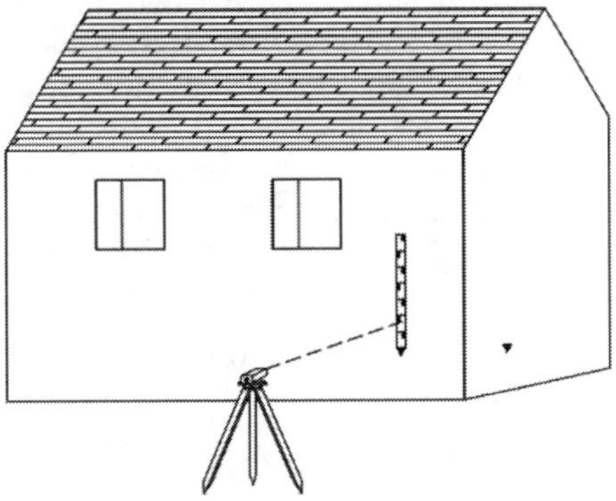

Fig. 1 Levelling measurement of benchmarks.

2.1.2 Direct measurements of deflections

It is often, however, impossible to stabilise observation points, and other measurement methods have to be employed such as the method of direct projection on the levelling rod (Fig. 2). Verticality measurement, consisting of rectilinear projection of characteristic points of each wall corner on the rule, next to the classical levelling of benchmarks, is the method most often used in practice by measurement and geological teams of mining companies. Wall verticality measurement can be obtained directly and rapidly using an electronic spirit level used in construction practice. It is quite problematic in this method to determine the application point, especially for building walls undergoing substantial deformations.

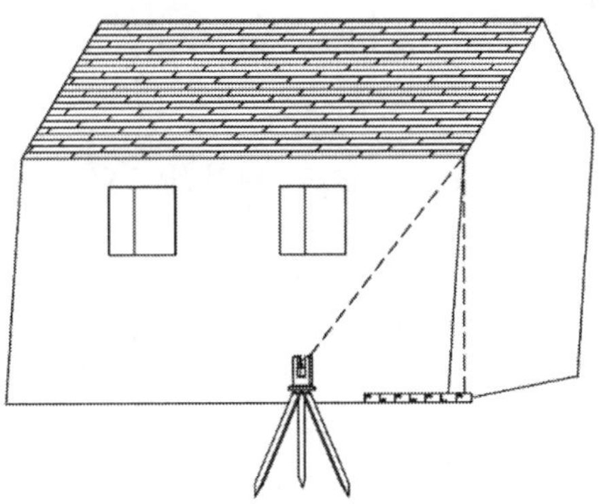

Fig. 2 Method of direct projection on the levelling rod.

2.2 Registration of floor sloping

The majority of methods of measuring the deflections of building walls is also applied for measuring floor sloping. The satellite method, where open space is required, is an exception here. An electronic or laser spirit level is widely used. In order to achieve greater accuracy, one can apply a levelling measurement of floors inside the building by placing a levelling rod on the floor, and for ceilings - by applying the rod to the ceiling (so-called measurement of inside diameter of the room).

2.3 Registration of other structural elements

For the rectification of multi-floor buildings, it is justified to perform measurements of elevator shafts (if such exist). Practical experience shows that an elevator core construction is made with the highest accuracy and is not subject to modifications during structure operation. Measurement results for floor slopes and wall surface deflections reveal large disparities between particular floors and apartments, mainly due to refurbishments made.
Elevator shaft deflection measurements can be made by means of an optical plummet, with the method of offsets from the measurement plumb line to the core walls. Deflection is measured from the plumb line of each core wall, with measuring at the same time the height of the level at which the measurement was made. Fig. 3 shown below lists results of independent measurements of core deflection in two elevator shafts.

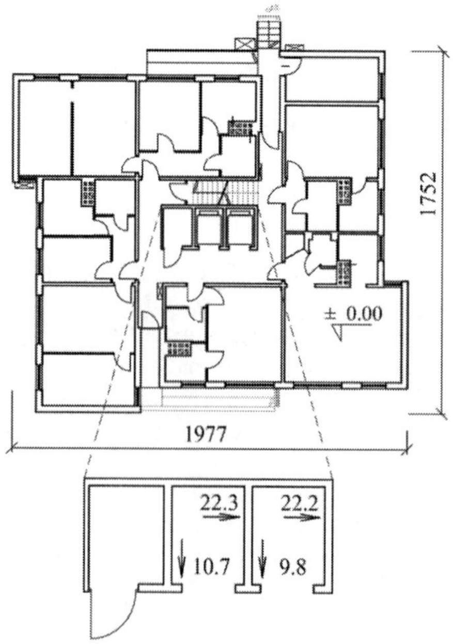

Fig. 3 Floor plan of the repetitive floor with the direction and value of the elevator shaft deflection marked.

Similarly, more modern laser plummets will also find their application here.
The interesting solutions given in [12] show the application of, e.g. digital cameras, sensors measuring length changes and inclinometers, or laser scanning when taking an inventory of mining shafts, which can also be successively used in surface conditions.

2.4 Spatial registration of structure (x,y,z)

Methods involving spatial measurement have to be often used for inventory measurements of structures, the purpose of which is to determine their technical condition. A full picture of the deformed structure can be obtained by measuring horizontal and vertical displacements. The most traditional techniques include, in particular, the method of space intersections and the polar method [13] (Fig. 4). Measurements carried out with modern equipment such as a tacheometer enable to make a measurement with appropriate accuracy.

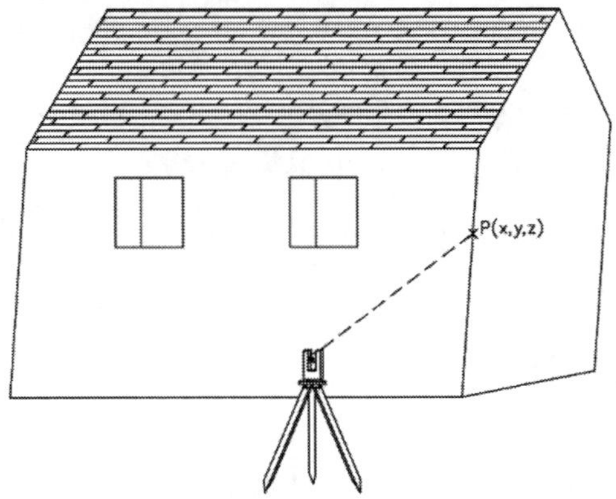

Fig. 4 Spatial registration of the structure (x,y,z)

The most modern methods, providing a full picture of the studied structure, include ground laser scanning. An automated tacheometric measurement enables to obtain an immense amount of data such as a cloud of points within short time. Data processing is most work intensive.

Satellite measurements with GNSS systems are also used in displacement measurements. The static method is used most often for ensuring appropriate accuracy, combined with classical methods, in order to make a reference to the global coordinate system. Proper visibility must be ensured for this method, i.e. the horizon cannot be covered, which is most problematic for this method.

The photogrammetrical method consisting of geometrical transformation of photographs is the least accurate method.

Relative displacements and deformations during continuous registration of structure behaviour can be obtained by using such sensors as clinometers, inclinometers, extensometers, etc., classified as telemetric methods.

3 Terrain sloping measurements

Land observations can be performed with commonly used geodesical methods.

Measurements of peripheral surveying lines, e.g. with levelling techniques, are performed in order to observe changes in terrain sloping around building structures. A direct terrain measurement, using electronic tachymetry or GPS techniques, allows to observe the occurring horizontal displacements. The photogrammetrical method (by preparing aerial photos or satellite images), aerial laser scanning (so-called LIDAR) or the radar interferometry method - InSAR, have to be applied to produce data for larger areas.

4 Examples of observations in scope of deflections of buildings structures and land sloping

In general, it is assumed that at the stage of forecasting the building deflections, a building deflection corresponds to terrain sloping [1]. Observation results often reveal, however, that differences exist between building deflection and land sloping, which is also presented in this chapter.

The studies were carried out on the land subject to intensive impact of underground mining exploitation. The observations covered the measurement of deflections of 12 buildings carried out with an electronic spirit level for corners of external and internal walls and floors. Fig. 5 below shows an example of measurements prepared in the form of a building view with the deflection values measured.

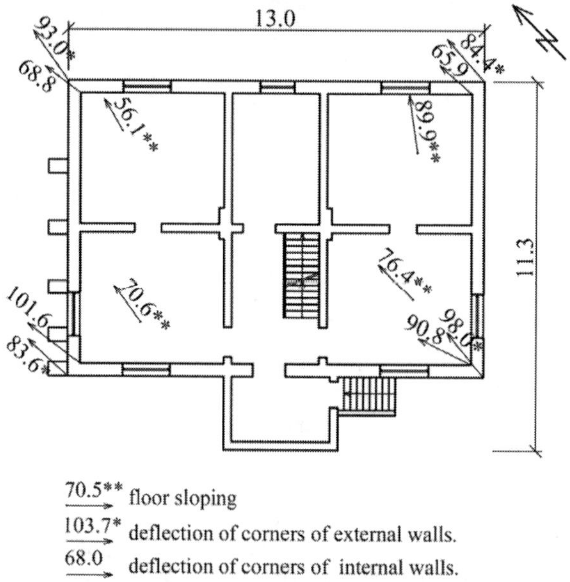

Fig. 5 Example of building view with the measured floor sloping and deflection of corners of external and internal walls.

The deflections measured and forecast terrain slopes are listed below.

Table 1. List of the measured building deflections with calculated terrain deformation indicators for mining exploitation performed.

No.	Average deflection [mm/m];			Terrain deformation indicators calculated with programmes of Prof. J. Białek for exploitation since building construction to 10.01.2017		
	Corners of external walls	Corners of internal walls	Floors	Sloping T [mm/m]	Horizontal deflection ε [mm/m]	Curvature radius R [km]
1.	59.2	61.5	54.2	47.6	-6.3	8.5
2.	77.0	72.9	74.3	43.9	-2.8	9.5
3.	60.5	-	-	29.9	-20.8	1.1

4.	63.3	73.6	72	65.6	77.4	-0.5
5.	71.3	77.5	83.9	31.8	-32.7	1.9
6.	115.2	111.2	91.1	90.7	-14.1	-2.1
7.	101	99.1	99.1	107.1	-37.9	1.3
8.	89.8	81.8	73.3	44.1	-41.4	0.9
9.	92.5	-	-	27.1	-31.5	1.7
10.	57.8	62.4	55.3	22.8	9.8	-5.6
11.	50.3	54.9	34.7	35.7	45.3	-1.2
12.	57.8	69.3	68.1	50.7	-7.3	-4.7

Differences in results of deflection measurements of inner corners relative to outer corners result principally from the renovations made inside the buildings and from the measurement error due to the fact that the point of applying a spirit level, especially on the strongly deformed walls, was not unambiguously identified. Differences in deflections of particular structural elements also prove the structure's spatial work. This phenomenon is especially visible on areas subject to intensive impacts such as compressive and tensile strains and terrain curvatures. The majority of structures show a higher average deflection than terrain sloping. Apart from the influence of underground mining exploitation, the differentiated morphology of the analysed terrain and the layers deposited in the subsoil in form of silty and sand formations and fine sands with washing ability should be indicated as the cause of the existing phenomenon.

5 Summary

The measurements of vertically deflected structures are carried out as part of inventory measurements for control and research purposes, to explain the impact of mining exploitation on the structure, and also when identifying the initial value in the rectification process. The measurement method applied should provide data with adequate accuracy. The buildings subject to mining impacts, especially in the form of deformations and curvatures, exhibit various types of deformations, e.g. form-related deformations of walls. Methods enabling to observe horizontal and vertical displacements should be employed to observe full changes in the coordinates x,y,z. The effect of non-mining factors, especially soil conditions and construction errors of the building structure itself, should be taken into account when interpreting the deflection values obtained.

References

1. M. Kawulok: Evaluation of functional properties of buildings considering mining impacts. Publishing House of the Building Research Institute, Warsaw (2000)
2. W. Mika, L. Muszyński: Assessment of admissible deflections of residential buildings on mining areas considering structural safety. Occupational Safety and Environmental Protection in Mining, 3, 10-15 (2011)
3. K. Gromysz, T. Niemiec: Selected issues in straightening the vertically deflected building structures. 3rd Scientific and Training Conference, Safety and protection of building structures on mining areas, Ustroń-Zawodzie. Scientific works of the Central Mining Institute (GiG), Katowice, 43-65 (2010)
4. PN-81/B-03020 - Foundation bases.
5. T. Niemiec, R. Zaborek: Accuracy in construction of civil structures and accuracy of their rectification. 2nd Conference: Building structures on mining areas. Katowice (2016)

6. T. Kowal: Proposed determination of the damage value in form of permanent vertical deflection of the building body. Mining Review, 10, 164-169 (2014)
7. M. Kawulok, L. Chomacki, B. Parkasiewicz, L. Słowik: Demolition of 25 residential buildings due to intensive impacts of mining exploitation. 26th Research and Technical Conference "Building Failures". Szczecin (2013)
8. M. Wyleżoł: Determination of measurable deflection of buildings in the aspect of their designing. Introduction to selected issues in civil engineering, scientific works of PhD students. Publishing House of the Silesian University of Technology, Gliwice, 191-198 (2017)
9. S. Szpetkowski: Deformation measurements on mining areas. Śląskie Publishing House, Katowice (1978)
10. W. Bochenek, Z. Motyka, H. Passia, A. Szade: Laser control of buildings and engineering structures subject to impact of underground mining in the region of the Upper Silesia Coal Basin. Scientific works of the Central Mining Institute (GiG), Katowice (2007)
11. J. Kwiatek at al.: Safety and protection of building structures on mining areas. Publishing House of the Central Mining Institution, Katowice (1997)
12. M. Jabłoński, W. Jaśkowski: Review of shaft pipe inventory-taking techniques. Construction and Architecture, 15(3), 63-74 (2016)
13. 2. J. Ćwierz, W. Cinal, A. Kampczyk: Measurement of building facade walls deflections from the vertical plane. Scientific Journals of the Częstochowa University of Technology, 22 (172), 48-65 (2016)
14. P. Gruchlik: Ground 3D laser scanning, experience and perspectives. Mining Review, 5, 20-25 (2015) [14]

Assessment of burned coal shale properties based on cyclic load

Marcin Grygierek[1,*], *Piotr Kalisz*[2], *Krzysztof Pacześniowski*[2], *Andrzej Pytlik*[2], *Magdalena Zięba*[2]

[1]Silesian University of Technology, Faculty of Civil Engineering, Gliwice, Poland

[2]Central Mining Institute, Katowice, Poland

Abstract. Road surfaces that are subjected to cyclic loads generated by vehicle wheels must meet the requirements concerning the durability in the assumed period of use. The durability of the layered pavement construction systems depends on the value and frequency of the load as well as on the mechanical features of its individual layers. Layers of unbound, mechanically stabilized mixtures are a significant aspect of surfaces that are susceptible. Mixtures of this type can be applied both to the subgrade layers as well as to the bottom pavement layers, including the improved course. Considering the cyclic nature of the load on the surface of the entire system, mechanically stabilized layers are subject to continuous, but slow, densification during the period of use, which results in the formation of permanent deformations and so-called structural ruts. Post-mining waste is frequently used in road construction. which is the so-called burned shale that can be used for the bottom layers of the surface and layers of the improved subgrade (soil replacement). This material was the subject of the analysis. The evaluation was based mainly on the results of pilot studies covering cyclic loads of the layer/course made of the so-called red shale. The applied research method was aimed at preliminary assessment of its suitability for the assessment of the behaviour of the disintegrated medium under the conditions of test loads simulating the movement of vehicles.

Keywords: static and cyclic plate load testing, burned coal shale

1 Introduction

Road surface constructions usually use layers of mechanically stabilized aggregate which is not chemically bonded. These layers can be applied in the lower layer of the subgrade and/or in the lower layers of the surface and in the improved course. Mainly these layers and ground course are responsible for the development of the so-called structural deformations that are observed in the wheel tracks. Structural deformations are also referred to as so-called structural ruts. Their origin is related to the densification of the granular

* Corresponding author: marcin.grygierek@polsl.pl

medium influenced by cyclic loading of layers generated by heavy vehicles. The evaluation of the properties of layers made of mechanically stabilized material is based on the evaluation of the modulus that a given material can achieve and the deformation as a function of the number of loads [1]. The assessment of the mechanical properties of aggregates, including the assessment of their non-linear behaviour, is the subject of many studies and analyses [2,3,4,5,6]. The value of the modulus is determined by several factors, including grain size of the mixture, its moisture, density of the layer prior observation, condition of the subgrade (its moisture content), deformations occurring in mining areas. Previous research and observations indicate that the modulus of the unbound layer can be further increased even still in the course of using the pavement, although this depends on the conditions occurring in subgrade of the analysed layer.

The article presents the results of pilot laboratory tests of aggregate subjected to cyclic loads, simulating the operation of aggregates during the use of the pavement. The research involved the use of burned shale, which is a waste created during the mining of hard coal. This material is applied to all layers of earthworks [7] to lower layers of pavement construction and layers of the improved subgrade.

2 Characteristic of burned coal shale

The use of waste materials in construction, especially communication systems, is extremely important in terms of the possibility of development, which reduces the problems associated with their storage and environmental impact [3]. Current rich experience obtained thanks to the use of waste accompanying mining of hard coal shows that burned coal shale, referred to as splinter shale, show significantly better properties than not burned coal shales. The predominance of burned shales results mainly from better resistance to water and frost and lower susceptibility to fragmentation during their compaction (Fig. 1).

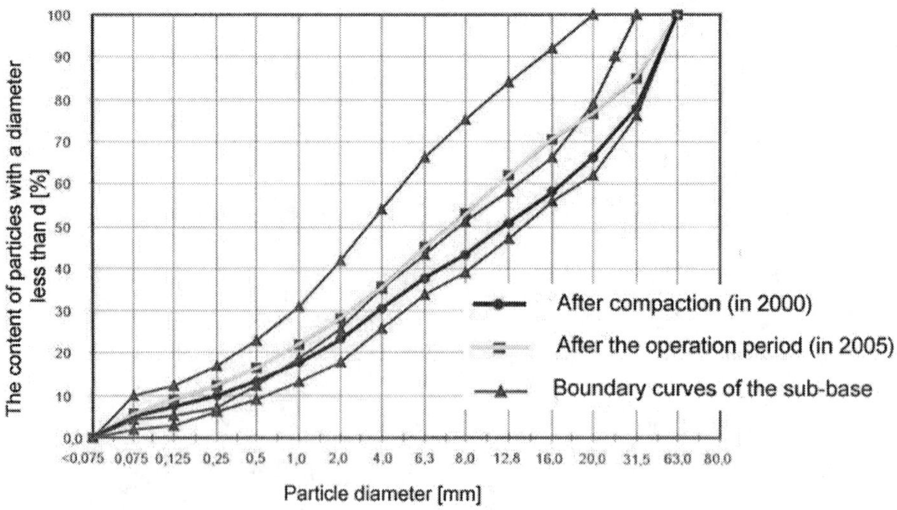

Fig. 1. Granularity of aggregate made of burned coal shale after compacting and five years of road use [8]

As Rafalski points out [8], burned coal shale can be used for road foundations and other lower layers. Rafalski [8] is one of the many researchers who presented the field experiments describing the behaviour of burned coal shale (Fig. 1) applied in a road. These works present such aspects as deformation moduli, determined on the basis of a static VSS test, in different periods of surface use, i.e. 0 – 5 years (Tab. 1) - after 5 years, it was observed that the parameters increased from 128 MPa to 142 MPa.

Table 1. Results of tests on deformation moduli of burned coal shale in road pavement [8]

Period of use	Secondary deformation modulus E_2 [MPa]	Deformation index I_0
0 months	128	2.0
3 months	129	2.0
6 months	136	1.9
5 years	142	2.0

3 Characteristics of the testing station

The tests of cyclic loading of aggregate were carried out in laboratory conditions, in a box with a cross-section shown in Figure 3 and a length of 1.20 m. The testing station was placed on a concrete stand, and the side walls were made of 2.5 cm thick steel sheet, which was additionally ribbed. An integral element of the station is the frame, which is a counterbalance to the hydraulic cylinder exerting a load on the surface of the layer system under test. Two layers of material were built in the box (Fig. 3).

The graining of the sand used indicates that it is a coarse sand with a uniformity coefficient Cu = 2.2, which classifies this sand as poorly graded (uniformly-graded) soil (Cu < 5), hardly compactable. The grain size of the shale was compared with the required field of proper graining for individual pavement layers according to Technical Requirements for Unbound Mixtures [6], as shown in Figure 2.

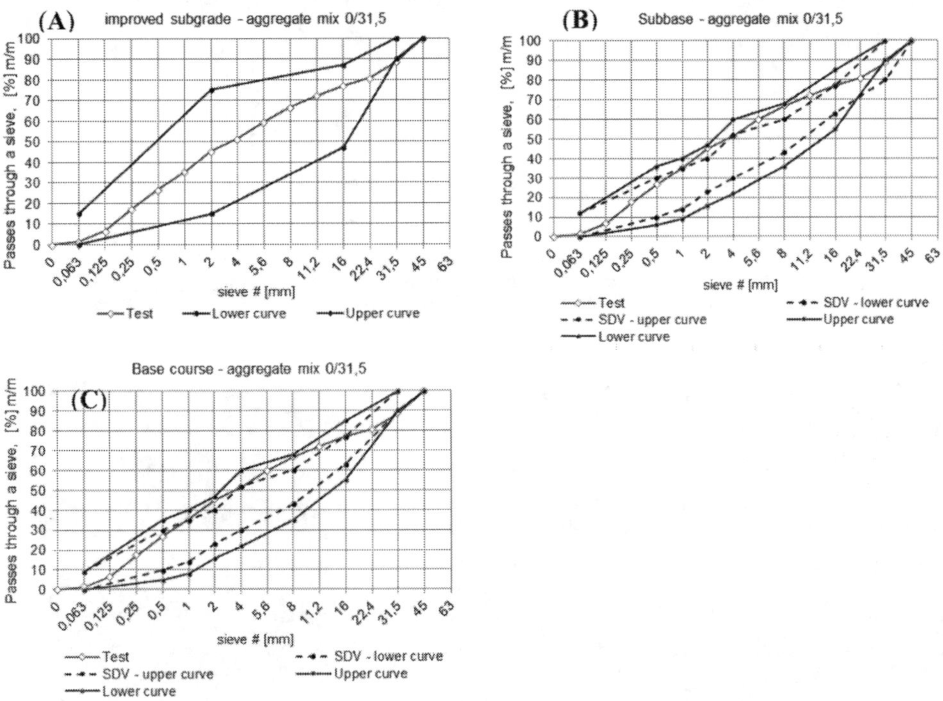

Fig. 2. Comparison of the graining curve of the tested mixture with proper graining curves [6]

The analysis of the course of the grain curve of the shale used clearly indicates that course is close to the upper limit of proper graining. In particular, it applies to areas of proper graining for subbase and subgrade. A comparison of the test results of selected aggregate properties and their comparison with the requirements for individual layers is presented in Table 2.

Table 2. Comparison of selected properties of burned coal shale and sand according to [9] and [6]

Subject material	Subject parameters		Test results	Requirements					
				Base course		Subbase	Frost protection layer		Layer of improved subgrade
				KR1-KR2	KR3-KR7	KR3-KR7	KR1-KR2	KR3-KR7	KR1-KR7
Burned coal shale	Sand index	[%]	**43.05**	≥ 45	≥ 45	≥ 40	≥ 35	≥ 35	≥ 35
	CBR index	[%]	**57.2**	≥ 60	≥ 80	≥ 60	≥ 25	≥ 35	≥ 20
	Compatibility of the grain size with a proper graining field		Fig. 2	yes*) 0/31.5	yes*) 0/31.5	yes*) 0/31.5	yes*) 0/31.5	yes*) 0/31.5	yes*) 0/31.5

*) graining curve close to the limit curve

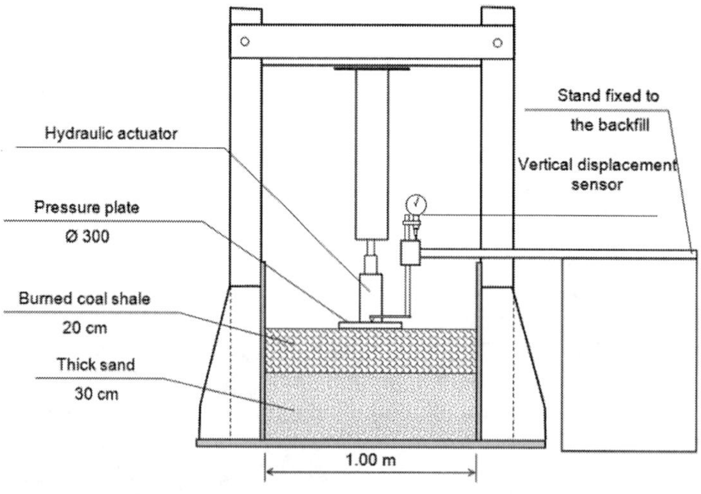

Fig. 3. Scheme of the testing station

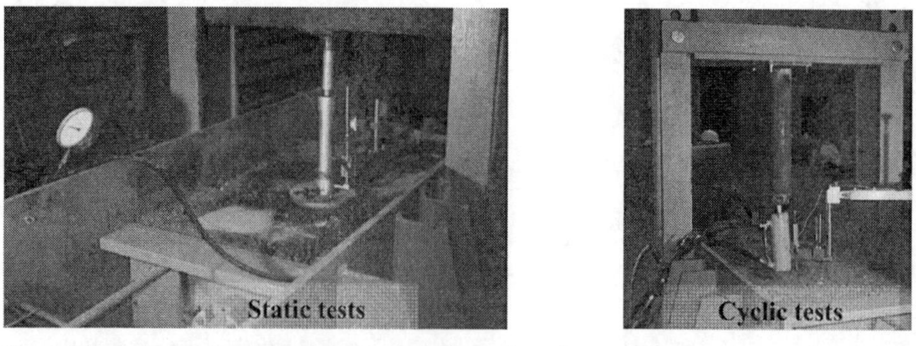

Fig. 4. The testing station during static and cyclic loading of aggregate

The static tests used the so-called VSS plate – i.e. a steel plate with a diameter of 30 cm, a hydraulic actuator and a stand for mounting the displacement sensor on a steel plate (Fig.4).

The tests based on cyclic loading of the aggregate layer were carried out by loading the surface of the system, also with a steel plate with a diameter of 30 cm, whereby the hydraulic actuator was controlled using a hydraulic pump and a control programme (Fig.4).

4 Implementation and tests

The tests were carried out in three stages. In the first stage, a static VSS plate measurement was made on the surface of the layer system (Fig. 5), marking the primary and secondary deformation moduli. In the second stage, the layer system was subjected to cyclic loading, using a hydraulic actuator, a hydraulic pump and a control programme. A load of 550 kPa with a frequency of 0.5 Hz was applied to the surface of the layers. In total, 430 load cycles were carried out (Fig. 7). The tests covered measurements of vertical displacements of the plate and the value of the force transmitted to the plate at a frequency of 25 Hz. The third stage of the research included the re-execution of the static measurement with the VSS plate (Fig. 6), which was used to calculate the primary and secondary deformation modulus. The final pressure was taken equal to that used during the cyclic test, i.e. 550 kPa.

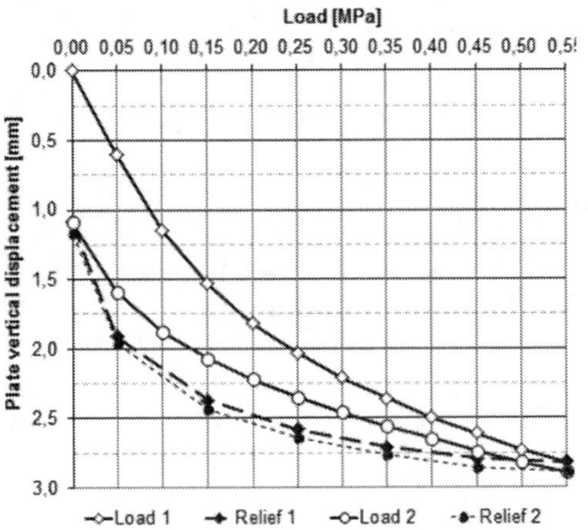

Fig. 5. Results of the static VSS plate measurement before the cyclic tests commence

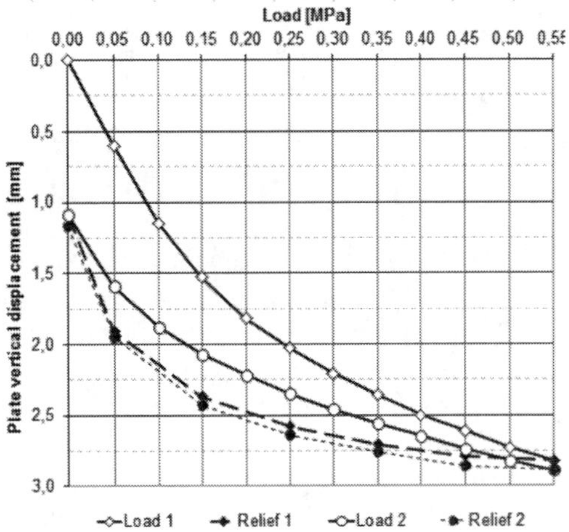

Fig. 6. Results of measurement with a static VSS plate after cyclic tests

The primary and secondary deformation moduli in the first and third stages were calculated as for the stress range of q = 150 – 250 kPa (Tab. 3).

Table 3. Calculated results of deformation moduli based on static tests (VSS plate)

Research stage	**Deformation modulus**		**Deformation indicator**
	E_1 [MPa]	**E_2 [M0Pa]**	**$I_0 = E_2 / E_1$**
Prior the cyclic loading test	36.3	75	2.06
After the cyclic loading test	45.0	80	1.78

In the second stage (cyclic test) the modulus was calculated in each load cycle according to the relationship (1). The results are shown in Figure 7.

$$E_{rN} = \omega \frac{\Delta q D}{\Delta s_N} \quad (1)$$

where:
ω – impact factor, for circular load surface ω = 1.0,
Δq – surface load increase [kPa],
D – diameter of the circular loading surface [m],
Δs_N – increase of surface displacement [mm].

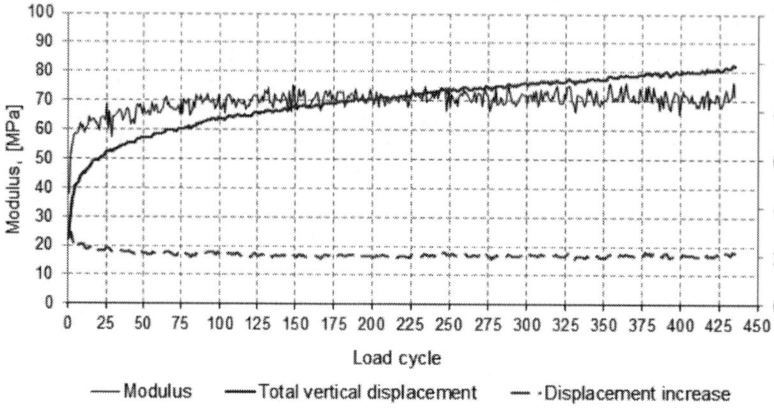

Fig. 7. Results of cyclic tests

5 Analysis and discussion of the results

The tested layer system before the cyclic test was characterized by a strain index $I_0 = 2.06$. This result indicates good density ($I_0 \leq 2.2$). The value of the secondary deformation modulus $E_2 = 75$ MPa should be described as sufficient, because the CBR index for this aggregate (CBR = 57.2 % - Tab. 2) suggests obtaining a relatively high stiffness for this layer. Similarly, when analysing the grain size of burned shale, inscribed in the field of proper graining for all surface layers, i.e. layers of improved subgrade, subbase and base course (Fig. 2), a high stiffness of this layer, higher than 100 MPa, is expected. Obtaining a modulus lower than expected should be combined with the stiffness of the lower layer, i.e. coarse sand, which is a uniformly-graded material, i.e. hardly compactable.

The results of cyclic loading of the layer system allow to observe a rapid increase of the vertical displacement (subsidence of the plate) in the initial cycles, and then slow stabilization of vertical displacements (Fig. 7) between subsequent load cycles. The vertical displacement of the plate, which decreases in subsequent cycles, indicates that the tested medium densifies and increases its stiffness (load-bearing capacity). It should be noted, however, that in cyclic tests the modulus began to stabilize at around 70 MPa. A similar value of the deformation modulus E2 = 80 MPa was obtained during the last (stage III) static test with the VSS plate. The obtained value of the modulus, at high density $I_0 = 1.78$, indicates a relatively small increase in stiffness of the upper layer, compared to the test carried out before the cyclic tests. This increase is definitely lower than expected. The unfavourable parameters of the sand layer constituting the subgrade of the examined shale should be considered as the probable cause of the relatively small increase of the secondary deformation modulus as well as the obtained value of the secondary deformation modulus whose value was lower than expected. These observations are also confirmed by the preliminary results of numerical analyses (Fig. 8), consisting in the identification of the parameters of the adopted Mohr-Coulomb model. The iterative method was used to find the parameter values, striving for the best fit of the theoretical curve (numerical calculations) presenting the displacement [mm] as a function of the load [kPa], to the experimental curve (from the VSS plate measurement). The performed back calculations indicate a low sand modulus (about 20 MPa) and a definitely higher value of the aggregate layer modulus (about 85 MPa) (Tab. 4). Taking into account such significant differences in resilient moduli, the properties of burned shale should be positively evaluated, because, as the tests indicate, the modulus of resilience of unbound layers is strongly determined by the stiffness of the base on which the subject layer is embedded. The results of numerical analyses also indicate one more phenomenon determining the value of the deformation modulus. This phenomenon is associated with displacing the material from under the load plate (Fig. 9), which causes an increase in the vertical displacement and reduction of the value of the modulus.

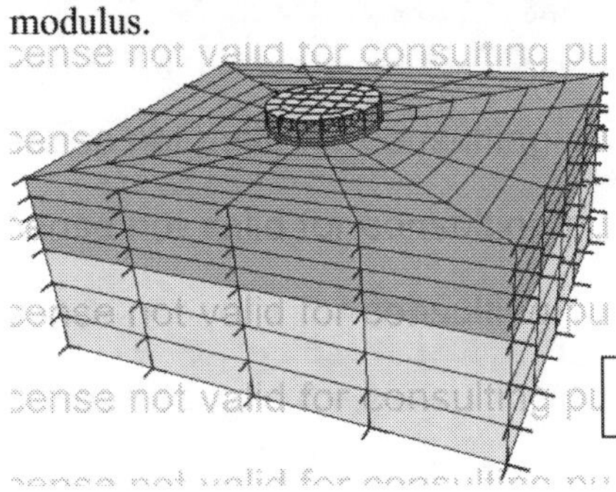

Fig. 8. Calculation model#

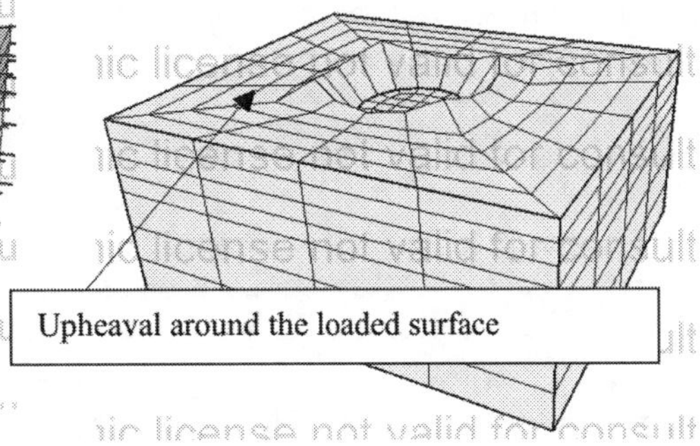

Fig. 9. Model deformations and calculated displacements#

Table 4. Parameters of the Mohr - Coulomb model

Layer	Iteration – 21.04					
	E [MPa]	c [kPa]	φ [°]	Ψ [°]	ν	γ [kN/m³]
Steel plate 2 cm	210,000	-	-	-	0.20	75
Burned shale 20 cm	85	2	45	15	0.35	17
Sand 30 cm	20	2	39	9	0.35	17

The fact that the grains of burned shale during the cyclic tests were broken (fragmentation) is not without significance for shaping the stiffness of the layer. This phenomenon is indicated by the experimental field studies described by Rafalski [8]. Therefore, this observation is significant as the grain curve of grain size of burned shale before the cyclic tests was carried out in close proximity to the upper curve of proper graining (Fig. 2 and Tab. 2).

6 Conclusion

The conducted tests using the station for cyclic loads of the system of unbound layers allowed in real conditions to assess the impact of repetitive loads on the mechanical properties of the tested layers. Such tests provide the opportunity to assess the behaviour of the layering system in conditions similar to those that occur during their use under a functional load, i.e. in conditions such as loading the pavement with circular motion.
The obtained results of the research well describe the causal relationship between the material properties of individual layers and the results obtained on the surface of this layer (deformation modulus). This type of testing station can be a tool for verification of other material solutions as well as it can be used to assess the impact of other phenomena occurring in the ground subgrade and/or in the pavement layers on its stiffness and

indirectly on its fatigue durability. The assessment of burned shale carried out on the basis of material tests (e.g. CBR, graining), cyclic loading and preliminary numerical analyses, it should be stated that this material meets the requirements for the frost protection layer and the lower layers of the pavement and layers of the improved subgrade. The results of the tests performed by other researchers, as well as the analysis of the results of own research, indicate the susceptibility of this material to fragmentation, which may affect the lower than expected increase in stiffness of the layer. In terms of the use of burned shale in the lower layers of the pavement, it seems justified to perform tests taking that would include variable moisture of the material, caused for example by periodic exposure to water, which impacts on the durability of road pavements [10]. Using the testing station will be also useful to evaluate the interaction between the geogrid and the unbound aggregate, which would broaden the research described by Grygierek and Kawalec [11].

References

1. D.J. White, K.R. Vennapusa Pavana, In situ resilient modulus for geogrid-stabilized aggregate layer: A case study using automated plate load testing, Transportation Geotechnics, 11, 120-132 (2017) http://dx.doi.org/10.1016/j.trgeo.2017.06.001
2. A. Hustvig, A new mechanistic model for prediction of permanent strain in unbound friction materials DOI: 10.13140/RG.2.1.4371.7528
3. J. Pieczyrak, Właściwości fizyczne oraz skład chemiczny i mineralny przepalonych odpadów kopalnianych (Physical Properties as well as Chemical and Mineral Composition of Burned Mining Waste) (in Polish), OTG, 62, 30-38 (1982)
4. Y. Qian, J. Han, S.K. Pokharel, R.L. Parsons, Determination of Resilient Modulus of Subgrade Using Cyclic Plate Loading Tests, Geo-Frontiers, ASCE, 4743-4751 (2011)
5. X. Sun, J. Han, J. Kwon, R.L. Rarsons, M.H. Wayne, Radial stresses and resilient deformations of geogrid-stabilized unpaved roads under cyclic plate loading tests, Geotextiles and Geomembrans, 43, 440-449 (2015)
6. WT-4, Unbound mixes for national roads, Technical Requirements. General Directorate of National Roads and Motorways (in Polish) (2010)
7. J. Pieczyrak, Characteristics and engineering usage of waste from coal mining, ACEE, 1 (2010)
8. L. Rafalski, Road base-courses (in Polish), Issue Studies and materials, 59 (2007)
9. Catalogue 2014-2, Catalogue of typical flexible and semi-rigid pavement structures. General Directorate of National Roads and Motorways (in Polish) (2014)
10. P. Rokitowski, M. Grygierek, Influence of high water contents on pavement layers stiffness caused by flooding, Civil Engineering Series, 16, 1 (2016) DOI 10.1515/tvsb-2016-0007
11. M. Grygierek, J. Kawalec, Selected Laboratory Research on Geogrid Impact on Stabilization of Unbound Aggregate Layer, Procedia Engineering, 189, 484-491 (2017) DOI: 10.1016/j.proeng.2017.05.078

Deformations of temporary wooden supports used to reduce building deflections in mining areas

Krzysztof Gromysz [1,*]

[1]Silesian University of Technology, Faculty of Civil Engineering, Gliwice, Poland

Abstract. Temporary supports, consisting of a stack of wooden elements and a hydraulic jack, are used in the process of removing deflections in buildings with one to three aboveground floors in mining areas. During uneven raising, the supports are loaded monotonically, unloaded and loaded cyclically. Laboratory tests were designed for the supports. For the investigated range of loads of 0 to 400 kN, under a growing load, a linear relationship exists between a load and the change in the stack length, which signifies that the deformations of wooden elements and displacements related to their mutual interactions increase proportionally. A seemingly higher stack stiffness is seen at the beginning of the unloading process and for cyclical loads, meaning that in this phase of loading, the material deformation of the wooden elements and the jack is responsible for changing the jack length in this load phase, with a negligible presence of mutual displacements of wooden elements. The support, after being unloaded, returns to the initial position and its permanent deformations are not observed. The stiffness of a temporary support decreases as the height of the stack of wooden elements increases.

Keywords: deflected buildings, temporary wooden support, stiffness

1 Introduction

Temporary supports, consisting of a hydraulic jack and a stack of wooden elements, are used in the process of removing buildings deflection in buildings with up to three aboveground floors. The use of such supports is common in the mining region of Silesia, where building deflections are widespread [1,2,3].

Building deflection is eliminated in several steps [1]. In the first step, piston hydraulic jacks are installed in openings made in underground floor walls. In the second step, force is excited in the jacks and the building is being torn. The tearing area is running through the load transmission points from the jack to the building (Fig. 1a). In the third step, the part of the building situated under the tearing area is elevated unevenly until a vertical position is reached. In this step, the object is revolving around an axis perpendicular to the direction of

* Corresponding author: krzysztof.gromysz@polsl.pl

the deflection being eliminated [4,5]. In the case where a given jack finishes piston shift range of 200 mm, it needs to be unloaded and underlaid. Oak wood elements with the side length of 300 mm and 400 mm and a height of 100 mm are used [2]. Supports have a different height because a building raising height. As a consequence, a temporary building support consists of a hydraulic jack positioned on a stack of oak elements with the height h_{wo} (Fig. 1b) or consists of a jack only. The wall fragments between the raised part of the building and the part remaining in the ground are filled up in the last step, and the temporary supports are removed. Because the displacements applied are cyclical and because it is necessary to underlay the jacks in the phase of uneven raising, three support loading phases can be distinguished: the longitudinal force value is increasing monotonically, the value is decreasing and cyclical variations occur in the longitudinal force value.

The stiffness of temporary supports needs to be known when designing the straightening of a building. In particular, stiffness in the longitudinal direction is the datum used in verification calculations and when programming the extension of pistons of particular jacks. It was found many times when straightening buildings that the stiffness is not a constant value. It is thus necessary to determine such stiffness in experimental tests.

A stand for testing the stiffness of supports is presented below and experimental studies are described the purpose of which was to determine stiffness in the longitudinal direction of the jack and a stack of wooden elements and a support as a whole.

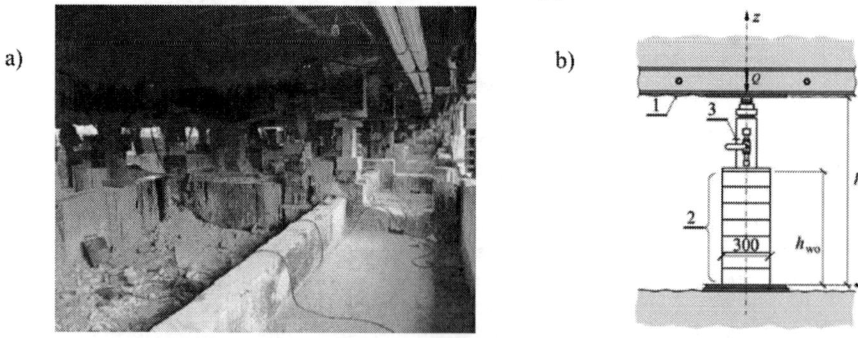

Fig. 1. Temporary supports during removal of building deflection; example of supports build in the building (a), schema of the support (b); 1 – building's tearing area, 2 – stack of wooden elements with the dimension of 300/400/100 mm, 3 – piston hydraulic jack

2 Tests project

During the removal of actual buildings deflections, temporary jacks are rested on steel sheets mounted in the concrete in the building wall part situated below the tearing area (Fig. 1). The support base cannot move or rotate then. The upper part of the support is a jack piston head provided with a hinge. A static scheme of a temporary support is therefore a jack with the height h, loaded with vertical force Q and horizontal force resulting from effects of the 2nd order or from the existence of loads in the horizontal direction (Fig. 2a). It is practically impossible to model such a scheme of a single temporary support in laboratory conditions, with the longitudinal force values Q reaching 400 kN. Another test scheme for a temporary support was used for this reason. It is a freely supported element with the height $2h$. A deformation axis of the element, considering the first buckling mode or the existence of load in the direction y, corresponds to a composition of an axis of two deflected supports working as a support (Fig. 2b). Under the influence of the 2nd order effects or as a result of loading with horizontal force, the section in the middle of the height of the system, with the height $2h$, is not rotated and corresponds to a support section of a support with the height h. The element being studied, with the height $2h$, is, therefore, a composition of two elements with the height h, and a centre of symmetry of the studied system is running through the centre of its span.

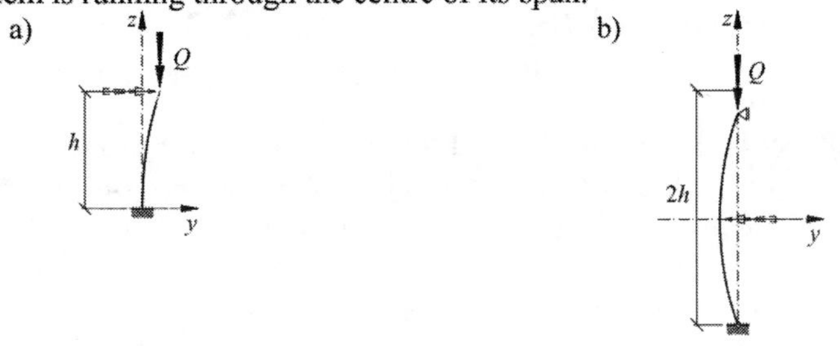

Fig. 2. Scheme of temporary support of the building with the height h (a), scheme of tested system with the height $2h$ (b); force acting in y direction was not considered in the tests

Laboratory tests were designed for a system with the height $2h$, set in a horizontal position. The system being studied consists of two piston jacks placed in the direction of the axis z, with their bases positioned towards each other, and of a stack of fourteen rectangular elements of oak wood situated between them (Fig. 3a). The length of each element in the direction x is 300 mm (Fig. 3b), in the direction y – 400 mm and 100 mm in the direction z. The direction of growth rings of all wooden elements coincides with the direction of the axis y. Steel plates with the dimension of 300/400/20 mm are situated between the bases of the jacks and a stack of wooden elements. The system was placed in a stand comprising two heads in the form of welded plate girders with a height of 440 mm. The heads were connected with two ties, and each tie consists of two sections [200. Dynamometers were mounted between the plate girders and the jack heads (Fig. 3a).

In the tests, the system was loaded with force with the value Q acting in the direction of the axis z, causing the system to be compressed. The force Q was caused by an active jack, and the jack situated on the opposite side was a passive jack (Fig. 3a). The following was recorded during the tests: the value of the force Q (Fig. 3b), variations in the length of the stack of fourteen rectangular wooden elements (u_{wo}) and variations in the passive jack length (u_{ja}).

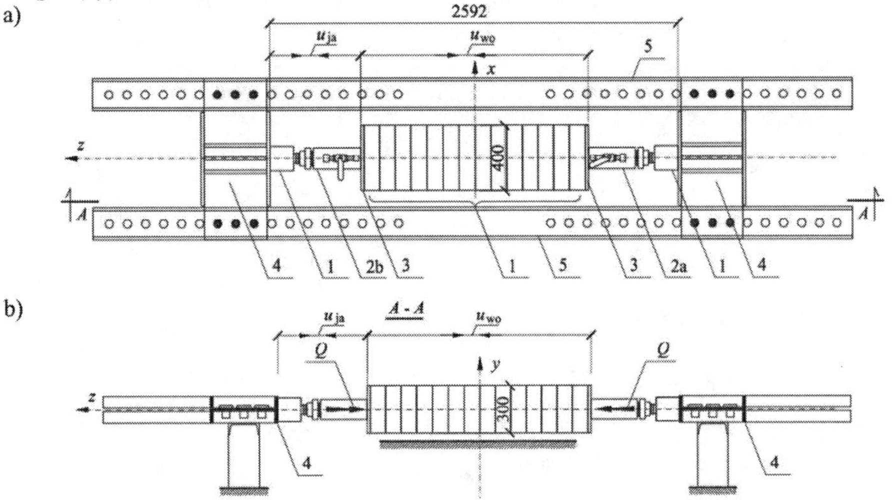

Fig. 3. Test stand and the tested system – the top view in the y axis direction (a), section $A – A$ (b): 1 – stack of fourteen rectangular wooden elements, each with dimensions of 300/400/100 mm, 2a – active hydraulic jack, 2b – inactive hydraulic jack, 3 – steel sheet with dimensions of 300/400/20 mm, 4 – the head of the stand, 5 – ties

3 Stiffness tests of the system elements

The adopted method of loading the system was reflecting the work of the actual support when straightening the building. For this reason, at first the value of the force Q was being increased monotonically from 0 to 400 kN. This part of the load was called phase A. After reaching the maximum deflection value, the value of the force Q was being decreased in such a way so that stiffness of the system can be determined with a decreasing load and with a cyclical load. System unloading from the force value 400 kN to zero was marked as the phase B of loading. Cyclical loads with the load amplitude of 50 kN with average forces of 375 kN, 325 kN, …, 25 kN were marked as the phase C of loading. For this reason, the system was being unloaded as follows. The value of the force Q was first decreased to 350 kN and then increased by 50 kN. Next, the force was decreased by 100 kN and increased by 50 kN. Variations in the value of the force Q during the test were as follows: 0 → 400 kN → 350 kN → 400 kN → 300 kN → 350 kN → 250 kN → 300 kN → 200 kN → 250 kN → 150 kN → 200 kN → 100 kN → 150 kN → 50 kN → 100 kN → 0 kN → 50 kN → 0. Table 1 lists loading phases with description of variations in the Q force value. The load was applied with a speed of 1 kN/s. The speed of load applying was small and didn't affect the rigidity of support.

A chart showing variations in the length u_{wo} of the stack of fourteen rectangular wooden elements under the influence of the changing value of the force Q, which was recorded every about 2 kN, is given in Fig. 4a. The phases of the monotonically growing load (phase A), falling load (phase B) and cyclical load (phase C) can be clearly distinguished in the figure. Figure 4b presents variations in the length u_{ja} of the passive jack depending on the value of the force Q.

Table 1. Load phases of the system

Load phase	Change of Q force value
A – monotonically growing	0 → 400 kN
B – falling	400 kN → 350 kN→ 350 kN → 250 kN→ 250 kN → 150 kN→ 150 kN → 50 kN → 50 kN → 0
C – cyclical	350 kN → 400 kN → 350 kN; 300 kN → 350 kN → 300 kN; 250 kN → 300 kN → 250 kN; 200 kN → 250 kN → 200 kN; 150 kN → 200 kN → 150 kN; 100 kN → 150 kN → 100 kN; 50kN → 100 kN → 50 kN;

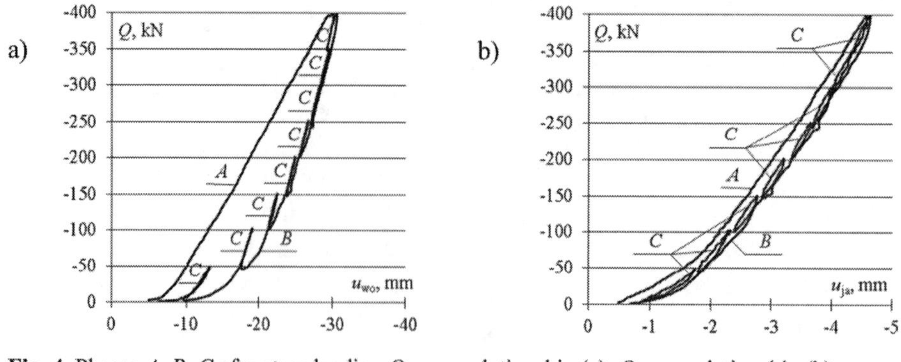

Fig. 4. Phases A, B, C of system loading $Q - u_{wo}$ relationship (a), $Q - u_{ja}$ relationship (b)

The following method was used of determination of the stiffness of a stack of wooden elements depending on the load phase. First, the polynomials were determined with the method of least squares describing the load – stack length variation relations for particular

https://doi.org/10.1051/e3sconf/20183603002

phases. Next, functions calculating the stiffness of stacks as the derivatives of such polynomials were calculated.

For a monotonic load (phase *A*), a satisfactory description of the $Q = Q(u_{wo})$ relationship was obtained with a polynomial of first degree. On the other hand, for the unload (phase *B*) and cyclical load (phase *C*), a satisfactory description of the $Q = Q(u_{wo})$ relationship was obtained with a polynomial of fourth degree. The values of the determination coefficient R^2 of all approximations were not lower than 0.99, denoting that the tests results were well described. The determined $Q(u_{wo})$ polynomials, for the stack of elements, with the corresponding values of R^2, are listed in Table 2. The polynomials describing the changes of $Q = Q(u_{wo})$ are presented graphically in Figure 5a. According to the adopted management procedure, the stiffness of a stack - corresponding to a given load phase - was determined as a polynomial derivative

$$k_{wo} = \frac{dQ}{du_{wo}} \tag{1}$$

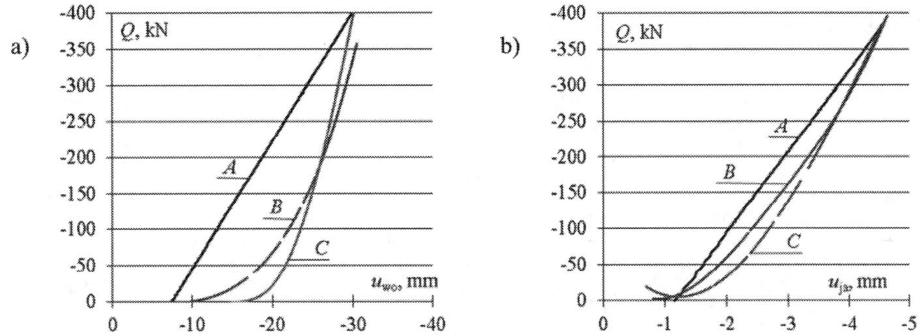

Fig. 5 Approximation of the force – displacement relationship – $Q(u_{wo})$ (a), $Q(u_{ja})$ (b)

For a monotonically growing load, stiffness of a 1400 mm high stack is described, with sufficient accuracy, by the constant value k_{wo} = 17.718 MN/m. For the unloading phase and cyclical loads, stack stiffness depends on the value of the loading force Q. As the value of this force is falling, stack stiffness decreases. For a maximum load of 400 kN, stack stiffness is about 50 MN/m. As the value of the loading force decreases, stack stiffness decreases to 3 MN/m with the force Q value of nearly zero. A decrease in stiffness, along with a decrease in the load, is higher for a cyclical load. Figure 6a shows a variation in stack stiffness for three types of load.

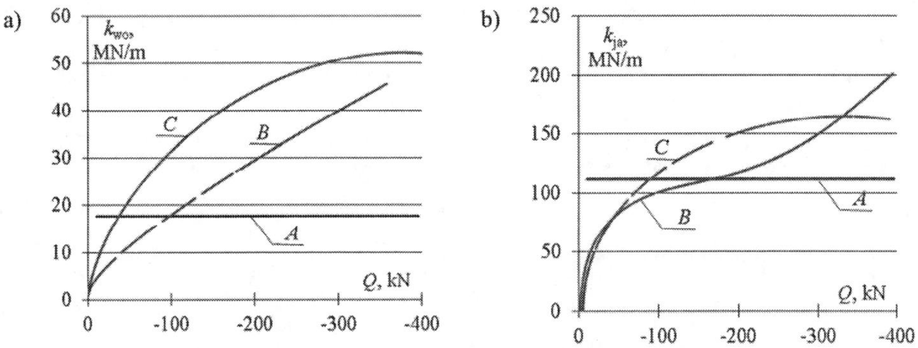

Fig. 6. Stiffness as a parameter depending on the loading force value and load phase – stiffness of stack of wooden elements (a), stiffness of hydraulic jack (b)

An analogous analysis was made for a variation in the length u_{ja} of a passive jack. The relations $Q = Q(u_{ja})$ were first described with polynomials. For a monotonic load (phase *A*), a satisfactory description was obtained with a polynomial of first degree. On the other hand, for the falling load (phase *B*) and cyclical load (phase *C*), a satisfactory description of the relationship $Q = Q(u_{ja})$ was obtained with a polynomial of fourth degree. The values of coefficients R^2 of all approximations were not lower than 0.98. The determined $Q(u_{ja})$ polynomials, with the corresponding values of R^2, are listed in Table 3. The polynomials describing the changes $Q = Q(u_{ja})$ are presented graphically in Figure 5b. The jack stiffness, corresponding to a given load phase, was determined as a polynomial derivative and listed in Table 3 and shown graphically in Figure 6b.

For a monotonically growing load, stiffness of a 1400 mm high stack is described, with sufficient accuracy, by the constant value $k_{wo} = 17.718$ MN/m. For the unloading phase and cyclical loads, stack stiffness depends on the value of the loading force Q. As the value of this force is falling, stack stiffness decreases. For a maximum load of 400 kN, stack stiffness is about 50 MN/m. As the value of the loading force decreases, stack stiffness decreases to 3 MN/m with the force Q value of nearly zero. A decrease in stiffness, along with a decrease in the load, is higher for a cyclical load. Figure 6a shows a variation in stack stiffness for three types of load.

Table 2. Approximation of the $Q - u_{wo}$ relationship and function describing the stiffness of the stack of rectangular wooden elements

Load type	Approximation of the load $Q = Q(u_{wo}); R^2$ Description of stiffness $k_{wo} = dQ / du_{wo}$
A – monotonically growing	$Q = 17,718u_{wo} + 129,87; R^2 = 0,999$ $k_{wo} = 17,718$
B – falling	$Q = -0,0006u_{wo}^4 - 0,0201u_{wo}^3 - 0,6955u_{wo}^2 - 9,023u_{wo} - 35,06; R^2 = 0,999$ $k_{wo} = 0,0016u_{wo}^3 + 0,0603u_{wo}^2 - 1,319u_{wo} - 9,023$
C – cyclical	$Q = -0,0009u_{wo}^4 + 0,815u_{wo}^3 + 24,963u_{wo}^2 + 321,55u_{wo} + 1503,3; R^2 = 0,9986$ $k_{wo} = 0,036u_{wo}^3 + 2,445u_{wo}^2 + 49,926u_{wo} + 321,55$

An analogous analysis was made for a variation in the length u_{ja} of a passive jack. The relations $Q = Q(u_{ja})$ were first described with polynomials. For a monotonic load (phase *A*), a satisfactory description was obtained with a polynomial of first degree. On the other hand, for the falling load (phase *B*) and cyclical load (phase *C*), a satisfactory description of the relationship $Q = Q(u_{ja})$ was obtained with a polynomial of fourth degree. The values of coefficients R^2 of all approximations were not lower than 0.98. The determined $Q(u_{ja})$ polynomials, with the corresponding values of R^2, are listed in Table 3. The polynomials describing the changes $Q = Q(u_{ja})$ are presented graphically in Figure 5b. The jack stiffness, corresponding to a given load phase, was determined as a polynomial derivative and listed in Table 3 and shown graphically in Figure 6b.

Table 3. Approximation of the $Q - u_{ja}$ relationship and function describing the stiffness of hydraulic jack

Load type	Approximation of the load $Q = Q(u_{ja}); R^2$ Description of stiffness $k_{ja} = dQ / du_{ja}$
A – monotonically growing	$Q = 111,68u_{ja} + 126,04; R^2 = 0,984$ $k_{ja} = 111,68$

B – falling	$Q = -2,6883U_{ja}^4 - 30,459u_{ja}^3 - 137,48u_{ja}^2 - 181,70u_{ja} - 74,90; R^2 = 0,999$ $k_{ja} = -10,7532u_{ja}^3 - 91,377u_{ja}^2 - 274,96u_{ja} - 181,70$
C – cyclical	$Q = 0,2661u_{ja}^4 - 2,4599u_{ja}^3 - 60,477u_{ja}^2 - 133,99u_{ja} - 83,953; R^2 = 0,999$ $k_{ja} = 1,0644u_{ja}^2 - 7,3797u_{ja}^2 - 120,954u_{ja} - 133,99$

For a monotonically growing load, jack stiffness is described, with sufficient accuracy, by the constant value k_{ja} = 111.68 MN/m. For the unloading phase and cyclical loads, jack stiffness depends on the value of the loading force Q. As the value of the force Q is falling, jack stiffness decreases. For a maximum load of 400 kN, its stiffness is about 150 MN/m, and as the load is falling, jack stiffness decreases to nearly zero.

4 Determination of stiffness of temporary support in longitudinal direction

The stiffness of the stack of wooden elements decreases along with its height. If the stiffness of the stack with a height of 1400 mm is known, the stiffness of the stack with the height h (Fig. 7a) can be determined from the following relationship:

$$k_{wo,h} = k_{wo}\frac{1400}{h} \tag{3}$$

The total variation in the length of a temporary support is a sum of variation in the length of displacements of the stack of wooden elements and the jack. As a result, a model of a temporary support is represented by a serial connection of the stack of wooden elements and the jack (Fig. 7b). Considering the above, the stiffness k of the whole support is determined from the following relationship:

$$k = \frac{k_{wo,h}k_{ja}}{k_{wo,h} + k_{ja}} \tag{4}$$

Figures 7c – 7d show charts presenting support stiffness in phases A, B, C depending on the height h_{wo} of the stack of wooden elements and for the value of the force Q = 100 kN, 200 kN, 300 kN.

a)

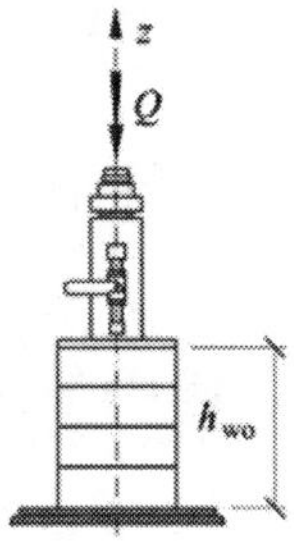

b)

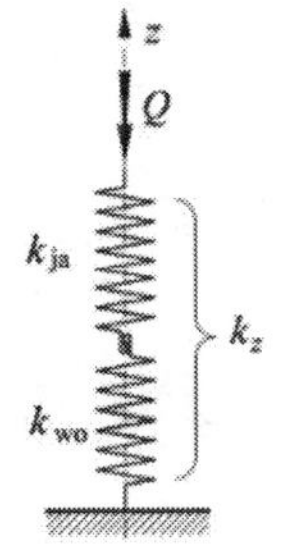

c)

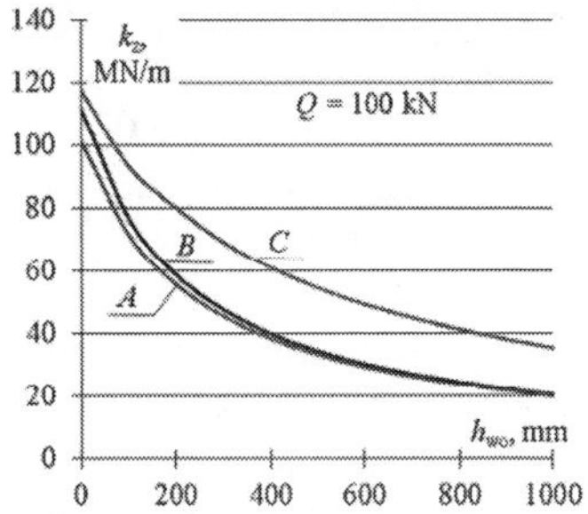

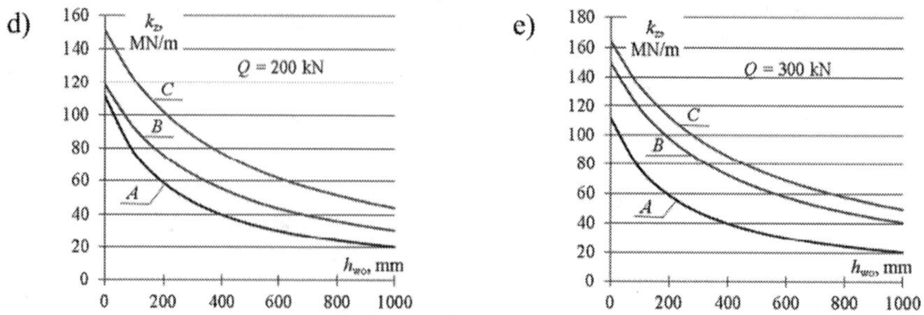

Fig. 7. Stiffness of temporary supports consisting of hydraulic jack and stack of wooden elements as depending on the height h_{wo} of the stack – sketch of the support (a), scheme of the support as serial connection of elements with stiffness k_{wo} and k_{ja} (b), stiffness of the support depending on the height h_{wo} of the stack for the load force Q (Q = 100 kN, 200 kN, 300 kN) (c, d, e)

The stiffness of a temporary support is decreasing along with the increased height of the stack of wooden elements. This stiffness is decreased by about three times if four wooden elements with the total height of 400 mm are installed. For the values of the load force of above 100 kN, support stiffness for cyclical loading and unloading is higher than for monotonical loading.

5 Conclusions

The stiffness of temporary supports in the longitudinal direction can be calculated as the stiffness of a system consisting of a serially connected jack and a stack of wooden elements. The stiffness of a temporary support is declining along with the increased height of the stack of wooden elements.

In the cycle of monotonic increase of the load, the support has the stiffness of 111 MN/m in the case where the support is represented by a jack itself. This stiffness is decreasing to 22 MN/m when a stack of the wooden elements mounted under the jack is 900 mm.

For a cyclical load with the maximum value of 400 kN, the support stiffness - in the absence of wooden elements - is approx. 160 MN/m, and with the height of the wooden elements stack of 900 mm, the stiffness decreases to 45 MN/m.

References

1. Gromysz K., Methods of Removing Buildings Deflection Used In Poland. IOP Conf. Ser.: Mater. Sci. Eng. 245 032096 (2017)
2. Gromysz K., Kowalski A., Mika W., Niemiec T., Repair of historical wooden church situated on mining area. Mining Review No. 3: p. 14-20 .(2015)
3. Kawulok M., Evaluation of functional properties of buildings considering mining impacts. Publishing House of the Building Research Institute, Warsaw (2000)
4. Kwiatek J. et al., Protection of building structures on mining areas. Publishing House of the Central Mining Institute, Katowice (1997)
5. Wyleżoł M., Determination of measurable deflection of buildings in the aspect of their designing. Collective work edited by K. Gromysz and R. Domagała. Outline of selected issues of Civil Engineering. Publishing House of the Silesian University of Technology: 191-198 (2017)
6. Gromysz K., Niemiec T., Selected issues in straightening the vertically deflected building structures". 3rd Scientific and Training Conference. Safety and protection of building structures on mining areas, Ustroń-Zawodzie, October 2010, Scientific works of the Central Mining Institute (GiG), Katowice: 43-65 (2010)

Application of the digital image correlation method in the study of cohesive coarse soil deformations

Janusz P. Kogut[1,*], and *Marcin* Tekieli[1]

[1]Cracow University of Technology, Faculty of Civil Engineering, Kraków, Poland

Abstract. Non-contact video measurement methods are used to extend the capabilities of standard measurement systems, based on strain gauges or accelerometers. In most cases, they are able to provide more accurate information about the material or construction being tested than traditional sensors, while maintaining a high resolution and measurement stability. With the use of optical methods, it is possible to generate a full field of displacement on the surface of the test sample. The displacement value is the basic (primary) value determined using optical methods, and it is possible to determine the size of the derivative in the form of a sample deformation. This paper presents the application of a non-contact optical method to investigate the deformation of coarse soil material. For this type of soil, it is particularly difficult to obtain basic strength parameters. The use of a non-contact optical method, followed by a digital image correlation (DIC) study of the sample obtained during the tests, effectively completes the description of the behaviour of this type of material.

Keywords: coarse soil, image correlation, unconfined compression

1 Introduction

It is difficult to identify the parameters to accurately describe soil. As is already known, soil is a mixture of solids, water, and gases. There are several methods to test soil. Namely, shear box tests and vane tests, as well as uniaxial and triaxial stress tests give some soil parameters, in a limited range, for different types of soils. In particular, they have caused problems for coarse and very coarse soils. Furthermore, it is difficult to extrapolate typical continuum models from coarse soils. Even after laboratory tests, it is necessary to implement the data into large, numerical scale models and then form a decision on whether the results are realistic, at least for a certain range of deformations (elastic, plastic). The last few decades have provided us with a tremendous opportunity to apply another class of the measurement methods, known as non-destructive tests. The internal structure of a soil sample may be examined for volumetric inspection with penetrating radiation (RT), such as

* Corresponding author: jkogut@pk.edu.pl

X-rays, neutrons, or gamma radiation. A combination of such tests, with the method applied in this paper, appears to be promising. Digital image correlation (DIC) generally refers to another class of non-contacting methods that acquire images of an object in digital form and perform image analysis to extract sensor-plane motions that can then be converted into full-field measurements on the corresponding object. One of the most commonly used approaches employs random patterns and compares subsets from initial, reference configuration and then the next configuration images to obtain a full-field of sensor-plane measurements. Fig. 1 presents the idea of DIC method [1].

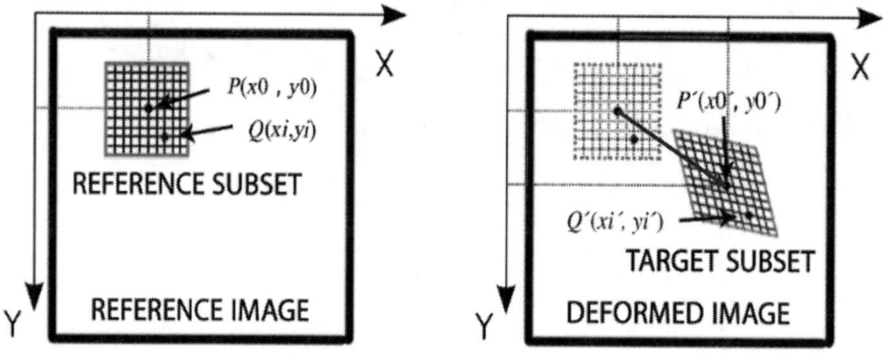

Fig. 1. The idea of DIC method – the reference image subset compared to the deformed target subset of points in the image processing [1].

The DIC method looks for the maximum of the correlation array between pixel intensity array subsets on several corresponding images, which gives the integer translational shift between them. It also estimates shifts in a finer resolution than the resolution of the original images, based on "subpixel" registration, because the measured shift is smaller than an integer pixel unit. An iterative approach may be used to maximize the interpolated correlation coefficient by using nonlinear optimization techniques. The two-dimensional discrete cross correlation r_{ij} is defined here [2-3] as

$$r_{ij} = \frac{\sum m \sum n \left[f_{(m+i,n+j)} - \bar{f} \right] \left[g_{(m,n)} - \bar{g} \right]}{\sqrt{\sum m \sum n \left[f_{(m,n)} - \bar{f} \right]^2 \sum m \sum n \left[g_{(m,n)} - \bar{g} \right]^2}} \tag{1}$$

where: $f_{(m,\ n)}$ is the pixel intensity or the gray-scale value at a point $(m,\ n)$ in the original image, $g_{(m,\ n)}$ is the gray-scale value at a point $(m,\ n)$ in the translated image, $\bar{f}$ and $\bar{g}$ are mean values of the intensity matrices f and g, respectively. In practical applications, one may use digital Fourier transformation (DFT) and then, by taking the complex conjugate and inverse Fourier transformation, receive the integer shift of $(\Delta x, \Delta y) = \arg_{(i,j)} \max \{r\}$. As for the deformation mapping, the mapping function that relates the images is derived by comparing a set of sub-window pairs over the whole images (fig. 1). The coordinates (grid points) namely (x_i, y_j) and (x_i^*, y_j^*) are related by the translations that occur between the two images. If the deformation is small and perpendicular to the optical axis of the camera, then the relation between (x_i, y_j) and (x_i^*, y_j^*) can be approximated by a 2D affine transformation such as:

$$x^{*} = x + u + \frac{\partial u}{\partial x}\Delta x + \frac{\partial u}{\partial y}\Delta y$$
$$y^{*} = y + v + \frac{\partial v}{\partial x}\Delta x + \frac{\partial v}{\partial y}\Delta y \quad (2)$$

where: u and v are translations of the centre of the subset in the X and Y directions, respectively. The distances from the centre of the subset to the point (x, y) are denoted by Δx and Δy. Thus, the correlation coefficient r_{ij} is a function of displacement components (u, v) and relevant displacement gradients [4].

2 Uniaxial unconfined compression soil tests

Three soil samples have been tested in order to get the Young's modulus E. The soil material is a quaternary colluvium formation deposit of the Carpathian flysch [5]. The uniaxial unconfined compression tests of the soil samples have been performed. Fig. 2 presents the experimental setup. The experiment has been performed with the application of Zwick 1455 static testing machine. It has a maximum load range of up to 20 *kN*, while the force measurement resolution is equal to 0.1 *N*, and the resolution of the vertical deformations measurements is equal to 0.002 *mm* [6-7]. The compression force and the vertical displacement has been measured. The camera used in the experiment is a Nikon DSLR D5300 (24Mpx, 6000x4000 px) with a Sigma 17-50 *mm* F/2.8 EX DC OS HSM lens.

Fig. 2. The experimental setup consisting of the high resolution camera, Zwick 1455 static testing machine (left), and the soil sample tested (right).

The resolution refers to the original value, which is obtained by means of an optical measurement, and, in this case, it is the displacement of a given point (marker). Thanks to the use of sub-pixel measurement, the measurement resolution was, depending on the test, from 0.01 to 0.02 px, yielding 0.0004 and 0.0008 *mm*, respectively. The collected image data has been processed using an optical measurement system (CivEng Vision) developed by the one of the authors of this paper [1,8-9].

3 Discussion of results

Fig. 3 presents the results of the deformations of the coarse soil sample in the solid state, in the following four virtual steps. It also may be visible, from the figure, that the virtual strain gauges were applied. The behaviour of the soil sample resembles the barrel shape with the vertical loading increasing, which may be seen when analysing steps #1 to #4. On the other hand, the sample cracks are noticeable at the bottom of left figure.

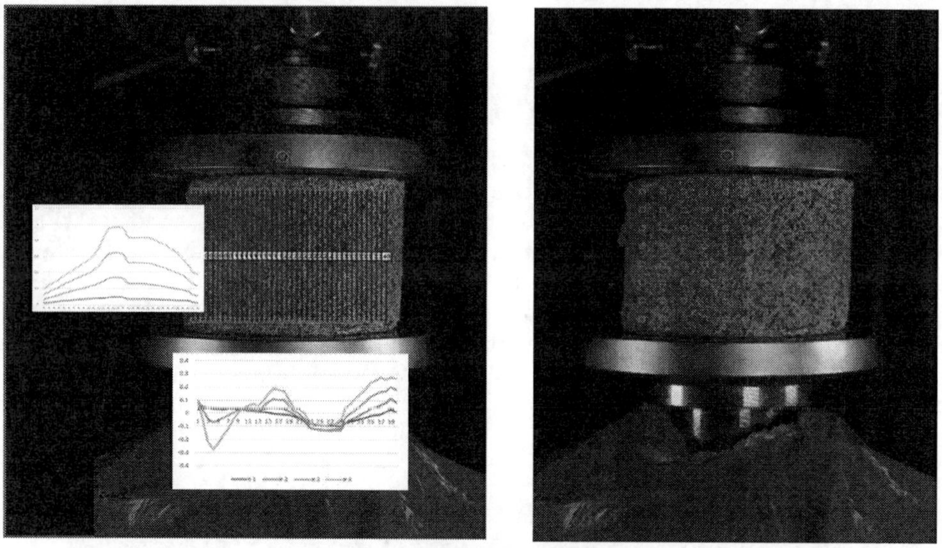

Fig. 3 The results of soil deformations in the four following steps of the compression test and the virtual strain gauges applied (left), and the virtual mesh grid (right).

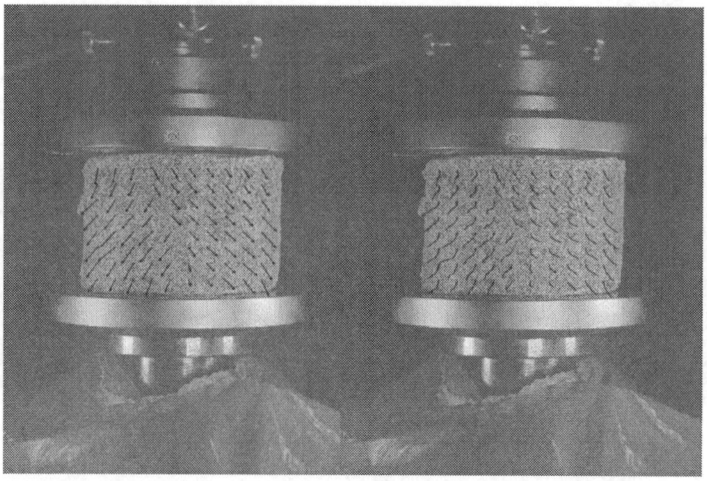

Fig. 4 The deformation vectors of the mesh grid (left), and the paths of the mesh points (right) put on the image of the soil sample.

Fig. 3 (right) presents the virtual mesh applied to the sample in order to obtain the real pathways of the mesh-points, chosen on the surface of the soil sample. The deformation vectors of the mesh grid (left) and the paths of the mesh points (right) put on the image of the soil sample are presented in Fig. 4. From that figure, one may observe the shear failure behaviour of the sample. The values are resized in order to notice this. It is even more apparent when colours are applied to illustrate the strains.

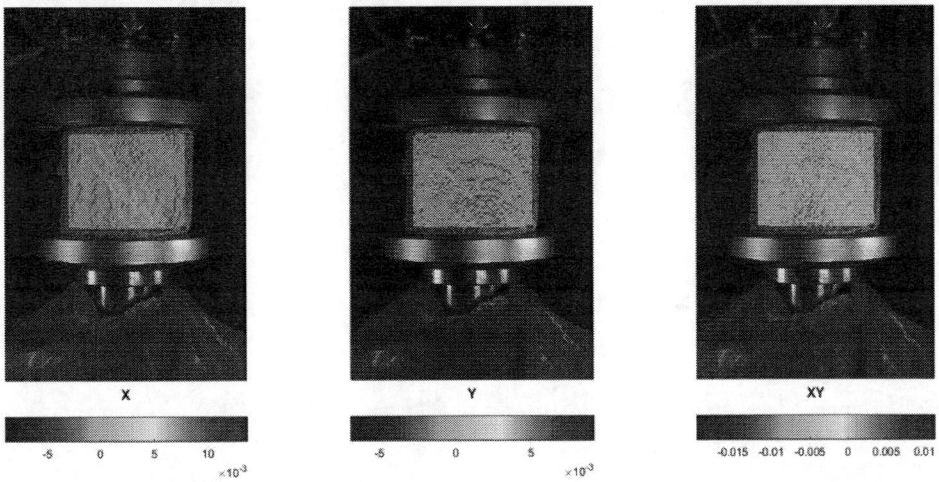

Fig. 5 The strain tensor components (ε_{xx}, ε_{yy}, ε_{xy}) digital maps at the early stage of the experiment, put on the image of the soil sample.

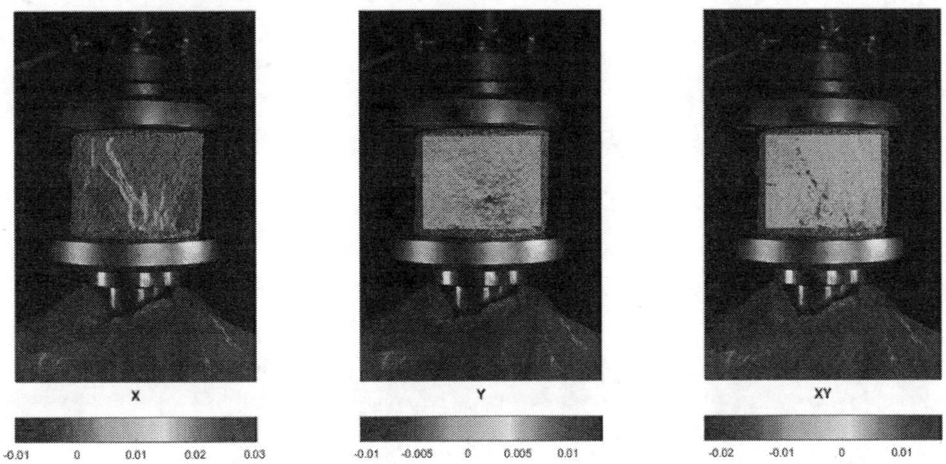

Fig. 6 The strain tensor components (ε_{xx}, ε_{yy}, ε_{xy}) digital maps during the medium phase of the experiment, put on the image of the soil sample.

Fig. 5 presents the digital maps of the strain tensor components, namely ε_{xx}, ε_{yy}, ε_{xy}, at the early stage of the experiment, put on the image of the soil sample. From the first tensor component ε_{xx} map, one may notice the beginning of the oblique crack, which is not visible in the other strain components, where the behaviour is chaotic. With an increase in force, the oblique crack is also visible in the tensor component ε_{xy} map, seen in Fig. 6. At the final phase of this experiment, another oblique crack is visible in Fig. 7, where the tensor component ε_{yy} map shows it, as well.

https://doi.org/10.1051/e3sconf/20183603003

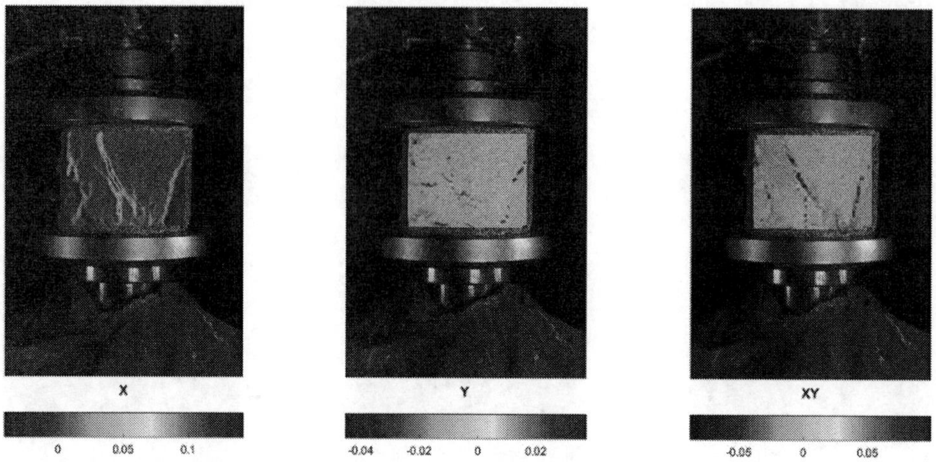

Fig. 7 The strain tensor components (ε_{xx}, ε_{yy}, ε_{xy}) digital maps at the end of the experiment, put on the image of the soil sample.

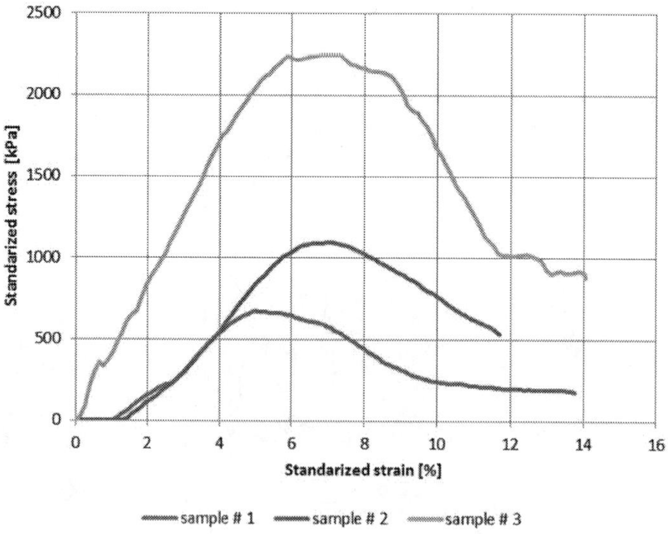

Fig. 8. Stress-strain curves, as received for the analysed soil samples.

Returning to Fig. 4, the first oblique crack is then visible after careful analysis of the vectors of the total deformations and the pathways of the mesh-points regularly chosen on the surface of the soil sample. Unfortunately, the second oblique crack is not visible in Fig. 4. As it was previously mentioned, the measurements of the compression loading and the vertical displacement has been performed during the test. Based on that data, Fig. 8 presents the stress-strain curves for all of the analysed samples. At the beginning of compression test for samples #1 and #2, the machine was adapting to the proper work, and thus, the stress is not visible, while displacements theoretically contributed more than 1% of strains.

Table 1. Young modulus and Poisson ration of the analysed soil samples.

Soil sample	Young's modulus *E* [MPa]	Poisson Ratio ν [-]	Soil density ρ [g/cm³]
# 1	23.6	0.26	2.18
# 2	29.8	0.35	2.19
# 3	43.4	0.34	2.17

Table 1 presents the Young's modulus *E* and Poison coefficient ν of the analysed samples in the range of stress-strains relation (ca. between 2% and 4%), as well as the density ρ of the soil samples.

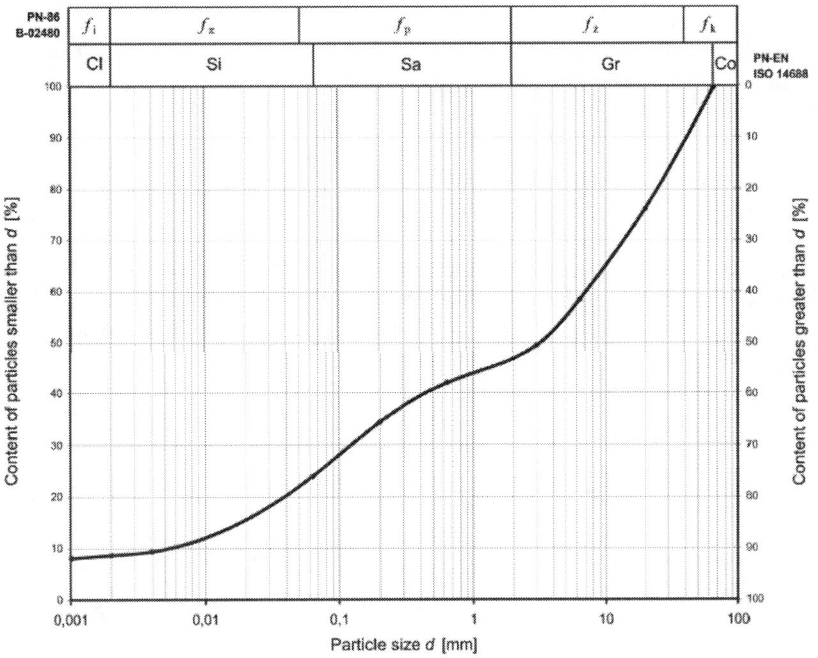

Fig. 9. Grain size distribution curve of the soil sample # 3.

The grain size distribution curve of the soil sample #3 is visible in Fig. 9 and the data from Table 1 may be helpful in the identification of the soil. According to the EN ISO 14688, the soil is classified as clayey gravel (clGr). However, bearing in mind the quality of that soil, the behaviour and the properties are those of cohesive clay rather than non-cohesive gravel.

Conclusions

The paper presented the application of a non-contact optical method for investigating the deformation of a cohesive coarse soil material. For this type of soil, it is especially difficult to obtain basic strength parameters. Three soil samples were tested. The uniaxial compression tests of the soil samples performed indicate that the behaviour of the soil is more similar to cohesive clay rather than to non-cohesive gravel. The use of a non-contact optical method, followed by the study of a digital image correlation (DIC) of the sample obtained during tests effectively completes the description of the behaviour of this type of soil material.

References

1. M. Tekieli, S. De Santis, G. de Felice, A. Kwiecień, and F. Roscini (2017). Application of Digital Image Correlation to composite reinforcements testing, Composite Structures, V. 160, pp. 670-688
2. F. Lagattu, J. Brillaud, and M-C. Lafarie-Frent (2004), High strain gradient measurements by using digital image correlation technique, Materials Characterization 53, pp. 17-28
3. L. Chevalier, S. Calloch, F. Hild, and Y. Marco, Digital image correlation used to analyze the multiaxial behavior of rubber-like materials, Eur. J. Mech. A/Solids 20 (2001), pp. 169–187
4. T.C. Chu, W.F. Ranson, M.A. Sutton and W.H. Peters (1985), Applications of Digital.Image-Correlation Techniques to Experimental Mechanics, Experimental Mechanics, pp. 232-244
5. T. Majcherczyk, Z. Pilecki, Z. Niedbalski, E. Pilecka, M. Blajer, and J. Pszonka (2012). The influence of geological engineering and geotechnical conditions on parameter selection of the primary lining of a road tunnel in Laliki. Mineral Resources Management 28 (1), pp. 103-124
6. A. Kwiecień (2014) Shear bond of composites-to-brick applied with highly deformable, in relation to resin epoxy, interface materials. Mater Struct 47:2005–2020
7. A. Kwiecień (2015) Strengthening of masonry using natural fibers bonding with highly deformable adhesives. GSTF J Eng Technol 3(2)
8. M. Tekieli, M. Słoński (2013). Application of Monte Carlo filter for computer vision-based Bayesian updating of finite element model, Mechanics and Control, 32 (4), pp. 171-177
9. D. Łątka, M. Tekieli. Optical measurements in the field of masonry construction laboratory tests. 2nd International Conference on Engineering Sciences and Technologies ESAT 2016, Tatranské Matliare, Slovak Republic

Results of the freeze resistance test, swelling index and coefficient of permeability of fine-grained mining waste reinforced with cements

Justyna Morman[1,*]

[1] Cracow University of Technology, Faculty of Civil Engineering, Krakow, Poland

Abstract. The article presents the result of laboratory tests for mining waste with grain size of 0 to 2 mm stabilized with cement. Used for stabilization of cement CEM I 42.5 R and blast furnace cement CEM III / A 42.5N - LH / HSR / NA and a plasticizer sealant. Cement was added to the mining waste test in the proportions of 5 - 8% in relation to the skeleton's weight. For the cemented samples, the freeze resistance test, swelling index, coefficient of permeability and pH of water leachate were tested. The addition of a cement binder resulted in diminishing the water permeability of mining waste and limiting the leaching of fine particles from the material.

Keywords: mining waste, mining waste with cements, the freeze resistance test,

1 Introduction

Extraction of mineral ores and energy resources results in mining waste. Since the 1980s, mining waste from various types of mines has been used to build earthworks for flood and road embankments [1, 2, 3, 4, 6, 7]. The mining waste from mines of the Upper Silesian Coal Basin have been stored on heaps for years, which is a big environmental problem. For example, each year, the Sobieski Mining Plant generates 4.6 million tons of coal and 2.2 million tons of mining waste. This waste contains mainly crumbs of claystone's, siltstones, clay minerals, shale and coal. The size of the mining waste distribution depends on the mining plant and the technology they use. Most often, this is waste with grain size from 0 - 200.0 mm, 0 - 125.0 mm, 0 - 31.5 mm, 0 - 2.0 mm and the so-called coal mills.

The article presents the results of freeze resistance tests, swelling and filtration coefficient which were carried out on mining waste from the Sobieski Mining Plant. Wastes with a grain size of 0 to 2 mm were selected for the test, they are not very resistant to physical and chemical ventilation. In the tests on the material, it was found that the material is difficult to compact to a density degree of compaction by 0.97 due to the fact that the material is hardly durable and under the effect of compacting it disintegrates into finer

* Corresponding author: jnmorman@gmail.com

fractions [2,4]. In order to improve its physical and mechanical properties, it was proposed to stabilize the material with cement.

In order to improve the physical and mechanical properties, it was proposed to stabilize the material with cement. Cement stabilization is also aimed at limiting the leaching of fine particles from the material when using material for the construction of flood embankments. The mining waste was stabilized with Portland cement CEM I 42.5 R and cement CEM III / A 42.5N - LH / HSR / NA and plasticizing sealant. Portland cement CEM I 42.5 R contains from 95 - 100% Portland clinker and up to 5% calcium sulphate binding time regulator. The main component of cement CEM III / A 42.5N - LH / HSR / NA is ground Portland clinker (35 ÷ 64%), granulated blast furnace slag (36 ÷ 65%) and calcium sulphate binding time regulator (up to 5%). It is characterized by low hydration heat (LH), high resistance to alkaline corrosion (NA) and high resistance to corrosive agents (HSR). For research purposes, also plasticizer sealing concrete and mortar were used to seal and waterproof Soudaproof IW. The Soudaproof IW sealing plasticizer is used for structural concrete which must withstand the external pressure of water, and for retaining walls, underground storeys, basements, foundations, patios, etc.

2 Preparation of trials, procedure, course and test results

The tests of freeze resistance, swelling and filtration coefficient were carried out for extractive wastes from the Sobieski Mining Plant with grain size of 0 to 2.0 mm. The mining waste consists mainly of clays and crushed-coal crumbs, the carbon content exceeds 10%, reaching in extreme cases even 20%. The waste contains up to 5% of sulphur [8,9]. In the grain size distribution test, about 18 - 23% silty and clay fraction and about 77 - 82% of sand fraction were determined. From the graining curve it follows that the waste is equiangular (U = 4.75), so it is a soil that is suitable for stabilization with cementitious cement in terms of grain size [10].

The tests of mining waste have been prepared for research:

- without added cements (ZG S1),
- with the addition of 8% Portland cement CEM I 42.5 R, with a diameter of 70.0 mm and a height of 35.0 mm (ZG S1 + 8% CEM I),
- with the addition of 8% Portland cement CEM I 42.5 R, dimensions 100.0 x 100.0 x 50.0 mm (ZG S1b + 8% CEM I),
- with the addition of 5% Portland cement CEM I 42.5 R and plasticizer Soudaproof IW in a proportion of 1.3 L / 100.0 kg of cement, 70.0 mm in diameter and 35.0 mm in height (ZG S1 + 5% CEM I + P),
- with the addition of 5% Portland cement CEM I 42.5 R and the Soudaproof IW plasticizer in a proportion of 1.3 L / 100.0 kg of cement, size 100.0 x 100.0 x 50.0 mm (ZG S1b + 5% CEM I + P),
- with the addition of 5% CEM III / A 42,5N - LH / HSR / NA metallurgical cement, 70.0 mm diameter and 35.0 mm height (ZG S1 + 5% CEM III),
- with the addition of 5% CEM III / A 42.5N - LH / HSR / NA, size 100.0 x 100.0 x 50.0 mm (ZG S1b + 5% CEM III)

The tests was concentrated in thin layers to the degree of compaction Is from 0.93 to 0.97, the moisture content in the range of 0.7 - 1.15 moisture content. The optimum moisture content for mining waste (ZG S1) was determined according [11] by method I, and it is 13.3%, for mining waste with cement 12.3%.

Attempts with cement additions were secured against moisture loss for 28 days of setting, then tested for frost resistance, swelling, filtration and samples of water leachate for testing pH.

2.1 Freeze resistance test

The freeze resistance test was carried out according to the [8] standard method - Simple method. After cementing, the samples were immersed in water for seven days. After saturation with water, the samples were frozen in air at -18 ° C for 6 hours. The samples were thawed in water at a temperature of ± 18 ° C for 3 hours. The freezing - thawing process was discontinued when clear cracks and cracks appeared on the sample or the percentage loss in mass was more than 5%.

Table 1 shows the weight of the test before and after the freeze resistance test. For mining waste with the addition of 8% Portland cement CEM I, in relation to the weight of the skeleton, during defrosting in water, the total disintegration of the molded sample was observed during the first cycle (SK 6, SK 7). Also two tests of mining waste with the addition of 5% CEM 1 cement and the Soudaproof IW plasticizer (SK 1, SK 2) disintegrated during the first cycle. From the SK 3 sample (ZG S1 + 5% CEM 1 + P), a weight loss of over 40% was recorded (figure 1). In the case of mining waste with the addition of 5% CEM III metallurgical cement, a decrease in the sample weight of over 5% was noted after the third cycle. The tests showed no scratches or cracks, only jagged edges were observed (figure 2).

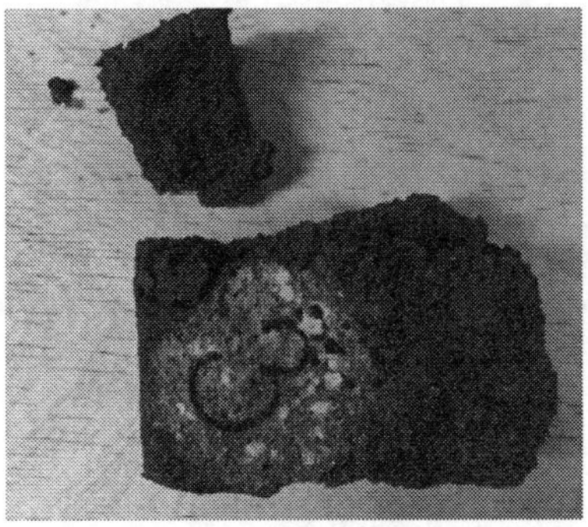

Fig 1. Mining waste sample with 5% CEM I cement + a plasticizer after the first cycles of freezing - thawing

Fig. 2. Sample of mining waste with the addition of 5% CEM lll cement before and after 3 cycles of freezing - thawing.

Table 1. Test of freeze resistance of mining waste reinforced with cements according to PN-B-06265 "Simple method".

No.	Addition of cement	Weight of the sample before the test [g]	Weight of the sample after the test [g]	Number of cycles of freezing - thawing	Loss of weight
SK6	ZG S1 + 8% CEM Ib	925.74	-	I	100%
SK7	ZG S1 + 8% CEM Ib	946.98	-	I	100%
SK1	ZG S1 + 5% CEM Ib + P	913.81	-	I	100%
SK2	ZG S1 + 5% CEM I b+ P	958.73	-	I	100%
SK3	ZG S1 + 5% CEM Ib + P	959.88	392.42	I	40.88%
SK4	ZG S1b + 5% CEM III	959,18	837,37	III	12,7%
SK5	ZG S1b + 5% CEM III	951,58	892,93	III	6.16%

2.2 Swelling index

The swelling process of mining waste and cement-reinforced mining waste was determined by the free swelling index Ep according to G. Olson [5], specifying the percentage increase in the height of the soil sample determined in the oedometer. The initial sample height is 35 mm. Figure 3 shows the swelling course for the samples tested. In the table 2 shows the swelling index for the trials after 10 and 60 minutes and after one day.

Table 2. Swelling ratio for mining waste with cement determined according to G.W. Olson [5]

	Swelling index Ep [%]		
Sample number	After 10 minutes	After 60 minutes	After 2500 minutes
Sample no. 1 (ZG S1)	-0.04	-0.21	-0.34
Sample no. 2 (ZG S1)	-0.02	-0.05	-0.06
Sample no. 3 (ZG S1 + 8% CEM I)	0.20	0.28	0.54
Sample no. 4 (ZG S1 + 8% CEM I)	0.15	0.46	0.81
Sample no. 5 (ZG S1 + 5% CEM I + P)	0.01	0.02	0.05
Sample no. 6 (ZG S1 + 5% CEM I + P)	0.004	0.02	0.05
Sample no. 7 (ZG S1 + 5% CEM III)	0.05	0.09	0.16
Sample no. 8 (ZG S1 + 5% CEM III)	0.004	0.007	0.02

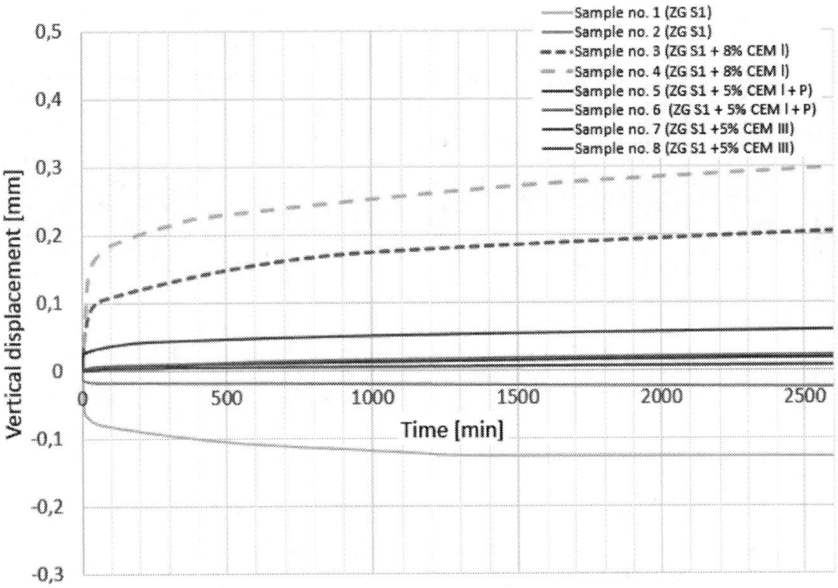

Fig.3 Swelling diagram of mining waste and cement reinforced extractive waste

According to the classific0ation of the free swell index Ep G.W. Olson, the swelling of the mixtures is very low. In the case of mining waste without additions of stabilizing substances, after the saturation of the sample with water, the sample height decreased.

2.3 Coefficient of permeability and pH

The results of the coefficient of permeability k for the mining waste test are presented in Table 3 and in figure 4. The test was carried out in an edometer according to [12] procedure. The measurement was carried out on pre-hydrated samples. Samples of mining waste with added cement showed a change in the coefficient of permeability with each subsequent measurement. The coefficient of permeability for mining waste ZG S1 without additions of binder is $4.3x10^{-4}$ m/s.

Extraction waste with the addition of 8% Portland cement CEM 1 was made for 9 measurements. The first three measurements showed a significant decrease in the coefficient of permeability, whilst subsequent measurements showed similar values, 5.70e-08 m/s, on average. In the case of the ZG S1 test with the addition of 5% Portland cement CEM 1 and the plasticizer in the proportion of 1.3L/100 kg cement, from 10 to 15 measurements were made for two independently cementing tests. The value of the coefficient of permeability changed successively from 1.6e-06 to $5.35x10^{-7}$ m/s. After stabilization, the coefficient of permeability for the ZG S1 sample with the addition of 5% Portland cement and the Soudaproof IW plasticizer is $5.57x10^{-7}$m/s, on average. Extraction waste tests with the addition of 5% CEM III cement also showed a decrease in the coefficient of permeability from $1.62x10^{-6}$ to $8.88x10^{-7}$ m/s, on average. The differences between samples 7 and 8 result from a different density index of the sample.

 https://doi.org/10.1051/e3sconf/20183603004

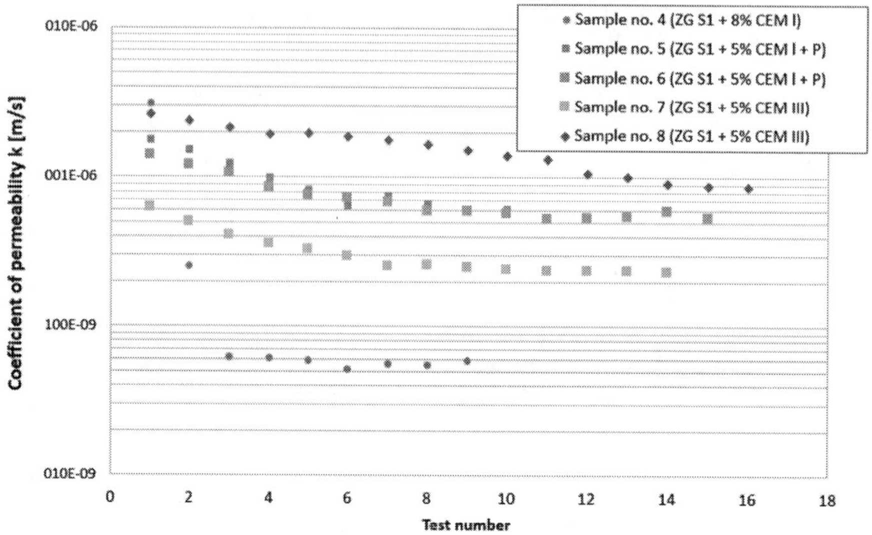

Fig.4 Coefficient of permeability k [m/s] for mining waste from the Sobieski Mining Plant with an addition of cements.

Table 3. Coefficient of permeability k [m/s] based on the procedure in the standard

Sample number	Coefficient of permeability k [m/s]		
	test No. 1	Last test	average value from the last 3 tests
ZG S1 + 8% CEM I	3.10e-06	5.88e-08	5.70e-08
ZG S1 + 5% CEM I + P	1.58e-06	5.35e-07	5.57e-07
ZG S1 + 5% CEM III	1.62e-06	5.52e-07	8.88e-07

Table 4 presents the results of the pH test of water flowing through the sample. In the case of the test with no added cement, it was observed that the water was cloudy and contained fine particles of the sample. The water effluent from the cement-addition trials was free of fine particles and had a more alkaline pH.

Table 4. PH of the water leachate.

Sample number	PH of the water leachate	Temperature [℃]
ZG S1	7.46	21.8
	7.49	22.5
ZG S1 + 8% CEM I	7.81	22.8
	7.92	21.9
ZG S1 + 5% CEM I + P	8.05	22.1
	8.03	21.7
ZG S1 + 5% CEM III	8.23	21.9
	8.34	22.6

3 Summary and conclusions

After the swell ratio test and the filtration coefficient for mining waste from the Sobieski Mining Plant with the grain size from 0 to 2.0 mm, the need to strengthen the research material with binders was identified. The main purpose of adding a cement binder was to reduce the leaching of material particles, reduce the water permeability and swelling, and increase the resistance to frost.

When testing the coefficient of permeability k for mining waste, a turbidity of water was observed. The test of pH of water leachate for samples without additives is 7.46 on average, and for mining waste with the addition of cements are more basic, e.g. with the addition of 5% CEM III cement - 8.34 (tab. 4). The pH of the leachate from the sample without cement additives is more acidic due to the presence of washed particles from the material. Environmental requirements according to the Regulation of the Minister of Environment of July 24, 2006 (OJ No 137, item 984, as amended) for all variants of the tested trials, where the pH range for water leachates ranges between 6.5 and 9.

The addition of a cement binder caused a reduction in the water permeability of the mining waste. In the case of cement tests, both CEM 1 and CEM III, a decrease in the filtration coefficient was observed together with the next measurement (figure 4). This is due to the swelling of extractive waste with the addition of cements and the reduction of pore spaces in the material. In the swelling diagram 1 and the filtration index 2, the relationship between the size and speed of the swelling increase to the decrease in the filtration index can be seen. The lowest coefficient of permeability k at the end of the swelling process was noted in the case of a mining waste test with the addition of 8% Portland cement CEM I and is on average the last three measurements 5.70e-08 m / s (poor water permeability of the material).

In the study of the freeze resistance test of mining waste with the addition of cements using the usual method according to PN-B-06265:1988, a noticeable improvement in strength in the case of mining waste with CEM III cement is noticeable. Samples with the addition of 5% CEM III / A 42.5N - LH / HSR / NA exceeded the permissible weight loss only after the third defrost. It should be noted that the maximum cement content in the sample is 5 - 8% in relation to the skeleton weight and moisture content in the range from 8.75 - 14.15%. The addition of a cement binder to mining waste from the Miners Sobieski Plant resulted in a reduction of leaching of fine particles from the material and reduced its water permeability.

Amongst the above-mentioned stabilization variants of the mining waste which was examined, the best results were obtained for the material stabilized with CEM III / A 42,5N - LH / HSR / NA cement.

References

1. J. Adamczyk, Basic geotechnical properties of mining and processing waste - a state of the art analysis, AGH Journal of Mining and Geoengineering, vol. 36, no. 2, 31 – 40 (2012)
2. Blight G. - Geotechnical Engineering for Mine Waste Storage Facilities. University of The Witwatersrand, Johannesburg, South Africa (2010)
3. A. Gruchot, Utilization of coal mining westes and fuel ashes for engineering purposes as a factor of environmental development and protection, Scientific Books of the University of Agriculture, no 410, Cracow (2016)
4. J. Morman, - Strengthening of mining waste with grain size up to 2 mm used in flood embankments (in polish), Contemporary construction in research of young scientists, University of Technology. Faculty of Civil and Environmental Engineering, Gdańsk (2017)
5. E. Myślińska, - Laboratory soil research. Ed. 2 changed and extended (in polish)., PWN Scientific Warsaw Publisher (1998)
6. E. Pilecka, J. Morman, - Utilization of fine-grained mining waste strengthened cement for the modernization of flood embankments (in polish), The Bulletin of The Mineral

and Energy Economy Research Institute of the Polish Academy of Sciences no. 101, 358-359, (2017)
7. E. Pilecka, J. Zięba - Preliminary studies of the dynamic stiffness modules of soil samples from the Solvay Sodium Plant waste landfill in Krakow, E3S Web of Conferences, 3rd International Conference on Applied Geophysics, Gniew, Poland (2017)
8. PN-B-06265. Concrete - part 1 Requirements, properties, production and abilities. Polish Committee for Standardization, Measurement and Quality. (2004)
9. Tauron – Wydobycie S.A., Development of a mining waste management program (PGOW) for PKW S.A. in Jaworzno – Sobieski Coal Mine (in polish),(2011)
10. PN-S-96012, Road car - Substructure and improved substrate with soil stabilized with cement. (1997)
11. PN-88/B-04481, Building land. Examination of soil samples. Polish Committee for Standardization, Measurement and Quality. (1988)
12. PKN-CEN ISOTS 17892-11, Geotechnical research. Laboratory soil research. Lot 11: Filtration testing with constant and variable hydraulic gradient. (2009)

Rockburst of parameters causing mining disasters in Mines of Upper Silesian Coal Basin

Renata Patyńska[1*], *Adam* Mirek[2], *Zbigniew* Burtan[3], *Elżbieta* Pilecka[4]

[1]Central Mining Institute, 40-166 Katowice, pl. Gwarków 1, Poland;

[2]State Mining Authority, 40-055 Katowice, ul. Poniatowskiego 31, Poland;

[3]AGH University of Science and Technology, 30-059 Kraków, al. Mickiewicza 30, Poland,

[4]Politechnika Krakowska , 31-155 Kraków, Warszawska 24, Poland

Abstract. In the years 2001-2015, 42 rockbursts were recorded in Polish coal mines. For the past 15 years the scale of the phenomena has been similar and ranges from 1 to 5 rockbursts per year. However, the number of recorded high energy seismic tremors of 10^8 and 10^9J (E) energy that has occurred in recent years, 2 to 5, is alarming. According to the data, 27 of tremors of E > 10^8 J energy that occurred between 2001 and 2015 caused 3 rockbursts. Confronting these data with seismic activity from 1989-2000, it should be noted that only 2 events out of 99 rockbursts caused tremors with energies of E>10^8 J. Against the background of the scale of seismic and rockburst hazards, the geological and mining conditions of the Upper Silesian Coal Basin (USCB) have been analysed, detailing the structural units in which the rockbursts occurred. On this basis, the author characterised factors that impacts on the mining excavations resulting in rockbursts that caused damage on a larger scale. These rockbursts had the characteristics of mining catastrophes and weak earthquakes not recorded in mining statistics of natural hazards of USCB so far.

1 Introduction

In the meaning of the applicable "Geological and Mining Law" [1] in Poland, "*rockmass is a discharge of spring energy accumulated in the rock, manifesting rock mass and acoustic phenomena.*" A particular case of tremor is a rockburst. It is defined as: "*a rapid discharge of elastic energy accumulated in rock mass, characterised by vibrations of rock mass carrying considerable energy, acoustic effects and a shockwave; this phenomenon causes destruction of rock structure in a roof, floor or bed with simultaneous dynamic rock relocation that fall into an excavation, and results in damages or destruction of the roof support or machinery and equipment.*"

* Corresponding author: rpatynska@gmail.com

 https://doi.org/10.1051/e3sconf/20183603005

In the Upper Silesian coal mines, over a hundred years of rock tremors have been occured. High-energy tremors are also felt on the surface, and even cause damage to cubature and linear structures.
The seismic hazard and the risk of rockbursts in Poland is currently occurring in most coal mines. Statistics [2,3] show that both the seismic and the rockburst risks increase year by year. It is conditioned by mining, geological, and technical and economic factors [4]. The most important of these are the natural factors, also referred to as geological. These include the initial rockmass pressure resulting from the deep mining (> 760 m), rock properties and rock mass tectonics (faults). Other factors largely impacting on the risk of rockburst in the roof are so-called tremor-prone layers, especially with high strength and thickness [5,6,7].
The next group are mining factors, which include, but are not limited to, the effects of exploitation (edges and residues) in neighbouring seams and limited operating conditions resulting from surface protection, residual deposit selection and "scarce" development work.
The aim of the analysis and research presented in the article was to indicate the mining and geological conditions characterising the areas where rockbursts have occurred in the last 15 years in the mines of the Upper Silesian Coal Basin (USCB).

2 General characteristics of rockubrsts in USCB mines

The rockburst hazard level in the USCB mines is illustrated by statistical data on the number of rockbursts recorded between 1949 and 2015 (Fig. 1) [2].
Between 1949 and 1962, 109 to 506 rockbursts yearly were recorded in the hard coal mining industry with output ranging from 78 to 104 million tonnes.
Since 1963, the number of rockbursts has dropped below 84 cases a year and continues to decline. Since 2009, production has been gradually decreasing, from 77.50 to 72.19 million tonnes in 2015. Data show that at output below 80 million tons, a comparable number of rockbursts, 1.57 every year.

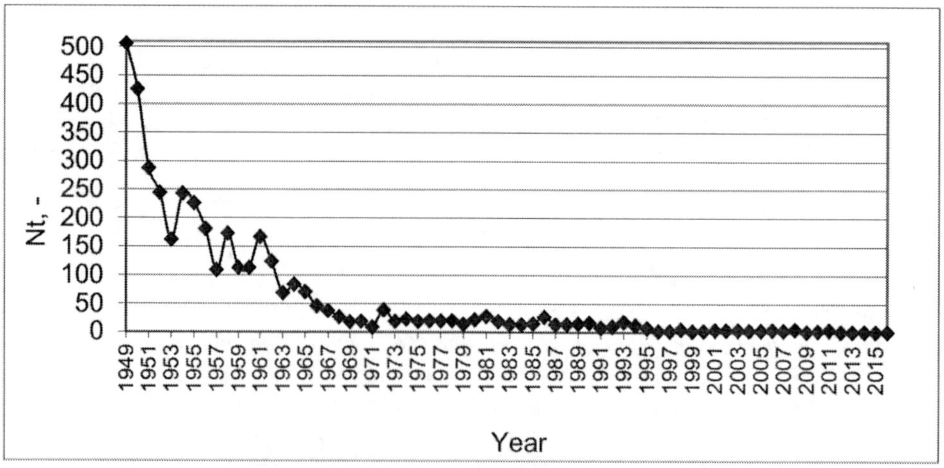

Fig. 1. Number of rock bursts (Nt, -) in USCB hard coal mines between 1949 and 2015 [2]

In spite of the more unfavourable mining and geological conditions in the vicinity of exploited seams, affected by past exploitation, the continuous development of mining technologies, mining experience and the use of effective prevention resulted in the occurrence of 1 to 5 rockbursts per year in the last 15 years (Fig. 2). It should be noted that only in 2008 the number of recorded seismic events amounted to 5.

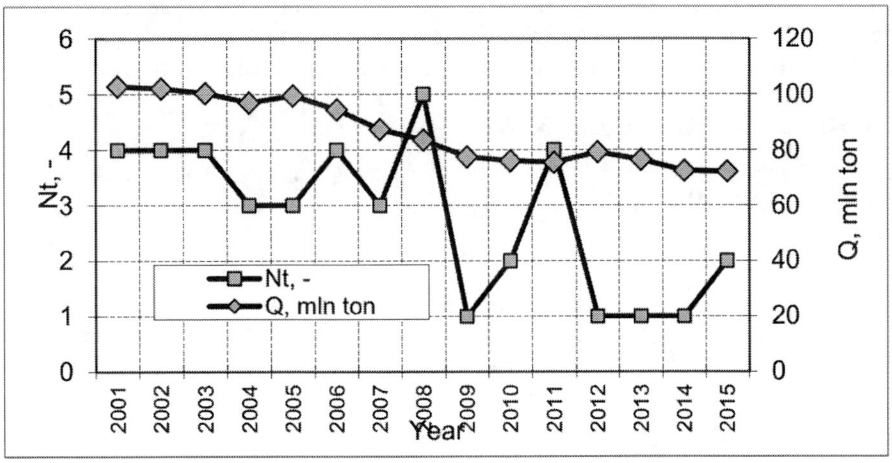

Fig. 2. Extraction of hard coal (Q, million tonnes) and number of rock bursts (Nt, -) in USCB mines in 2001–2015 [2]

The analysis of seismic activity distribution for individual USCB structural units indicates that the tremors are generated by mining operations and are located in five regions: Bytom Trough (NB), Main Saddle (SG), Main Trough (NG), Kazimierz Trough (NK), and Jejkowice and Chwałowice Troughs (NJ) [3].

It should be emphasized that between 2001 and 2015, the rockbursts were recorded only in three regions, namely Bytom Trough (NB), Main Saddle (SG) and Jejkowice and Chwałowice Troughs (NJ) [8]).

The characteristics of the regions indicate that seismic events that has led to rockbursts occur during the exploitation of deposits located at great depths, in the vicinity of fault zones, among the compact rocks. With the increase in the depth of the longwall operation and the preparatory work, there is a systematic deterioration of geological and mining conditions. In connection with the drawing process of coal resources, the mines more and more often decide to conduct exploitation of deposits in areas with residues, edges and/or in the area of former exploitation.

3 General characteristics of USCB seismicity

Between 2001 and 2015, the level of seismic activity was changing and depended on both the intensity of exploitation (total output) and the variability of the lithological and tectonic rock formations of the USCB.

Seismic activity was accompanied by the operation currently being conducted in five USCB regions (i.e.: NB, SG, NG, NK, and NJ) and is very diverse, ranging from tremors not detected by humans to events with characteristic of small earthquakes [9,10].

The most active seismic areas in the USCB are: Kłodnica Fault region with associated faults in Main Saddle (SG) and Main Trough (NG) area.

Generally, the number of tremors with highest levels of energy, 10^7 J, 10^8 J and 10^9 J has increased in recent years. The average number of tremors of 10^7 J energy between 2001 to 2013 amounted to 17, whereas in the 2014 up to 64 tremors of the same as mentioned energy level occurred.

Over the whole 15-year period, recorded tremors included one or two tremors of 10^8 J energy each year, with the exception of 2011, three such events, and in 2014 with up to 5 tremors. One tremor of 10^9 J energy occurred each year between 2008 and 2010, and in 2015, two [8].

The depth of exploitation is an important factor affecting the increase in seismic hazard especially after 2014. The data in Table 1 show that as the mean depth of the areas where the rockbursts are recorded increases steadily, the average energy of tremors that caused rockbursts increases. This particularly refers to tremors with energies of $> 10^8$ J, for which the depth of the rockburst ranged from 980 m to 1,069 m.

Table 1. Layout of the number of rockbursts against tremors that occurred between 2001 and 2015 according to energy level

Tremor energy level E, J	10^5	10^6	10^7	10^8	10^9
No. of tremors, Nw, -	14,824	2,801	333	23	4
No. of rock bursts Nt, -	10	12	17	2	1
The average energy of a tremor that caused a rockburst Ea, J	$6.3 \cdot 10^5$	$4.0 \cdot 10^6$	$5.0 \cdot 10^7$	$1.5 \cdot 10^8$	$4.0 \cdot 10^9$
Average depth of rockbursts Ht, m	858	830	813	980	1,069

The above data show that the significant contribution of the number of rockbursts as compared to the number of tremors have events of energy level $>10^7$ J.

However, high energy seismic tremors, 10^8 and 10^9J, are alarming, which occur more frequently than one in a year and have been recorded in the number of 2 to 5 tremors per year since 2006. The collected data indicate that 27 seismic events of $>10^8$ J, recorded between 2001 and 2015, caused 3 rockbursts.

4 Geological and mining factors determining the occurrence of rockbursts

The recipes of Geological and Mining Law [1] stipulate that seams classified as the 1st, 2nd, or 3rd level of rockburst hazard are deposited "*in rock mass and rocks prone to rockburst.*" Basically, in the conditions of tectonically not disturbed seams and rock masses, the level of rockburst hazard is determined by the stresses, the mechanical properties of the seam and the surrounding rocks and the seismicity generated by mining works.

The latter is the result of the damages of supported rocky rock layers with high strength parameters, and in particular the **tremor-prone layers**. They are thick, monolithic layers of sandstone and/or clay, of sufficiently high strength and thickness.

The risk of rockburst increases with the depth of the mining operation and the resulting increased stress in the rock and its gradient. [11,12,13].

In view of the above, it can also be assumed that with general low stress, rockbursts will occur at relatively high levels of energy of rock mass tremors.

The stresses in the rock can significantly vary within the range of the impact for gobs and the reach of the impact imposed by ***edges***, depending also on the vertical and horizontal distances from the excavation. Particularly unfavourable, due to the risk of rockburst, is summed up influence of the edges. Where the edge (seam, residue) is defined as the boundary line of the seam and gobs.

Accumulated impact of edges and/or **residue** (large surfaces), including the effects of compacting faults, creates particularly unfavourable stress and deformation conditions in the rock mass, and thus increased rockburst hazard. The **residue** is defined as a shallow part of the seam surrounded by goafs, in a shape of an island, peninsula or headland, with a width of 10 m to 0.3 H_r, which can concentrate stress (where: H_r - depth of deposited residue).

In the seams, the residue most often occurs near **faults** that exceed the thickness of the exploited coal bed.

Particular cases of residue are **support pillars** (protective), usually cut by numerous excavations, that often lead to excessive seam development. The pillar of carbonaceous charcoal along the gobs is separated by a pavement parallel to the gobs located in the distance of 10 m to 50 m. The support pillar, however, is a solid pillar with a width of 10-50 m and located between parallel pavements.

The geological and mining factors mentioned in the mining regulations that affect the risk of rockburst include:

- faults,
- tremor-prone layers,
- edges,
- residue,
- support pillars,

and excessive seam development, particularly adversely affecting at greater depths ($\geq$ 550m).

These factors represent a real threat of rockburst in active or developed mining excavations. Summarizes 42 rockbursts recorded in the years 2001-2015, grouped according to USCB structural units, the geological and geological factors characterizing the rock mass, and the location of recorded effects [2].

Data obtained in the location of the rockbursts indicate that these events occurred only in three regions of the USCB: Main Saddle (SG), Bytom Trough (NB), and Jejkowice and Chwałowice Troughs (NJ). The rockbursts that occurred in each of the area were conditioned by the interdependence of Nt = 5 of geological and mining factors.

Three of the events were registered in Main Saddle, energy reached $9 \cdot 10^7$ - $4 \cdot 10^9$ J and were recorded at the exploitation level 910-1,069 m. One event was registered in Bytom Trough, energy reached $2 \cdot 10^5$ J at 930 m. Jejkowice and Chwałowice Troughs were impacted by two rockbursts (Nt = 5), energy levels reached $9 \cdot 10^7$ J and $1 \cdot 10^8$ J at 1,050 m. (Table 2)

The most characteristic in terms of occurrence in the USCB rocky conditions are rockbursts caused by seismic tremors in operating conditions of the up to N_t=5 geological and mining factors. Although the tremors energy is > $2.00 \cdot 10^5$ J, but their vast majority, as many as 5 rockbursts, are related to seismic shocks with energy of $E_t = 9 \cdot 10^7$ - $4 \cdot 10^9$ J.

It should once again be stressed that the average shock energy of $7.47 \cdot 10^8$ J was related to rockbursts that occurred in the areas with the maximum number of geological and mining factors recorded. Such data translate into the "strength" of the impact of the identified 5 factors, expressed as the causes of rockbursts, and their effect on the extent of the effects of destruction and/or collapse of excavations [2].

Table 2 shows that 10 rockbursts, out of 42 analysed cases, occurred with the tremor energy of 10^5 J and were determined by 3, 4 and 5 Nt respectively. The depth of underground mining operation (gallery and longwall) in the areas where the effects of these rockbursts occurred was between 550 m and 1,150 m.

Table 2. Conditions of selected of tremors occurred in the USCB coal mines between 2001 and 2015

Item	Mine/date	Seam	Tremor energy E_t, J	Average seam depth H_a, m	Geological factors N_{geol}	Mining factors N_{gor}	No. of N_t factors; geological (N_{geol})/and mining (N_{gor})
USCB Structural unit		Main Saddle (SG)					
1	Wujek Ruch Śląsk, 18 Apr 2015	409	$4{\cdot}10^9$	1,069	-faults	-edges, -residue, -support pillars, -excessive seam development	1/4
2	Pokój, 27 July 2006	502	$9{\cdot}10^7$	970	-tremor-prone layers, -faults	-edges, -residue, -excessive seam exploitation	2/3
3	Halemba, 22 Feb 2006	506	$2{\cdot}10^8$	910	-tremor-prone layers, -faults	-residue, support -pillars, -excessive seam exploitation	2/3
USCB Structural unit		Bytom Trough (NB)					
4	ZG Bytom II, 26 June 2001	510	$2{\cdot}10^5$	930	-tremor-prone layers, -faults,	-edges, -residue, -support pillars	2/3
USCB Structural unit		Jejkowice and Chwałowice Trough (NJ)					
5	Rydułtowy Anna, 13 May 2006	703/1	$1{\cdot}10^8$	1,050	-tremor-prone layers, -faults	-edges, -residue, -excessive seam exploitation	2/3
6	Rydułtowy -Anna, 19 Dec 2006	703/1	$9{\cdot}10^7$	1,050	-tremor-prone layers, -faults	-edges, -residue, -excessive seam exploitation	2/3

To summarize the above, it should be noted that in case of mining co-occurrence of five geological and mining factors, mentioned as the primary causes of rockbursts, it is absolutely necessary to expect high energy tremors of $>10^8$ J which, as a consequence, can cause mining disasters.

Interestingly, in no case of rockbursts, there have been 6 factors at once (Table 2 and 4).

By analysing the above, it should be mentioned that the concept of rockbursts with features of mining disaster is a phenomenon of high energy shock, which results in, among others, in the form of fatal accidents, collective and/or severe injuries among miners and complete destruction of significant lengths of mining excavations.

Table 4. The number of rockbursts, grouped according to energy level (E_t, J) of seismic event and the number (N_t, -) of geological and mining factors

E_t / N_t	10^5	10^6	10^7	10^8	10^9
6 factors	-	-	-	-	-
5 factors / 5 rockbursts	-	-	**2**	**2**	**1**
4 factors / 22 rockbursts	-	**7**	**13**	-	-
3 factors / 15 rockbursts	**10**	**5**	-	-	-

This type of incident occurred in the case of three rockbursts recorded in the USCB mines in 2006 and 2015, which included, among others: fatal (2-4) and severe accidents (2-7), and extensive damage and destruction of excavations (with a reach of >500 m) [1] (Table 5).

Table 5. Geological and mining conditions of rockbursts in USCB coal mines with features of mining disasters

USCB Structural unit	Rockburst data	Depth, m	Fatal accidents /other	Length of destroyed/damaged excavations, m	Tremor energy Et, J	Geological and mining factors of a rockburst
Main Saddle (SG)	Wujek Ruch Śląsk, 18 Apr 2015	1,069	2/-	1200/-	$4 \cdot 10^9$	-faults, -edges, -residue, -support pillars, -excessive seam exploitation
	Halemba, 22 Feb 2006	910	-/2	Not determined >500	$2 \cdot 10^8$	-tremor-prone -layers, -faults -residue, -support pillars, -excessive seam exploitation
Jejkowice and Chwałowice Trough (NJ)	Rydułtowy Anna, 13 May 2006	1,050	4/7	-/300	$1 \cdot 10^8$	-tremor-prone -layers, -faults, -edges, -residue, -excessive seam exploitation

4 Conclusion

Based on seismic statistics in hard coal mines in Poland, it has been proven that such phenomena have started to occur in Polish hard coal mines only recently. The first of them was recorded in 2006, next in 2010, 2012 and in 2015.

Geological and mining conditions of rockbursts in USCB coal mines with tremor energy of $>10^7$ J causing mining disaster are shown in Table 5.

A detailed analysis of rockbursts with tremor energy of $>10^7$ J indicates the coincidence of 4 and 5 N_t factors, including mainly one geological factor (N_{geol}), fault, and two mining factors (N_{gor}), i.e.: residue and excessive development or exploitation of a seam.
This indicates the specificity of the current mining operation, most often below 900 m, under the influence of fault zones located in the areas of former exploitation, including in particular the excessive exploitation of a seam (Table 5).

References

1. Geological and Mining Law. J.L. 2011 No. 163 item 981, Act dated 9 June 2011.
2. R. Patyńska i in., Rockburst hazard. [in] Raport roczny o stanie zagrożeń naturalnych i technicznych w górnictwie węgla kamiennego. Central Mining Institute, Katowice (2002-2016)
3. K. Stec i in., Seismic hazards. [in]: Raport roczny o stanie zagrożeń naturalnych i technicznych w górnictwie węgla kamiennego. Central Mining Institute, Katowice (2002-2016)
4. I. Skrzypczak, A. Sobkowiak, W. Kokoszka, J. Kogut. The impact of underground mining on the threat of building structures. 17th International Multidisciplinary Scientific Geoconference - SGEM 2017, Proceedings vol 17, issue 11, Science Technologies in Geology, Exploration and Mining, pp. 623-630, ISSN 1214-2704, Albena, Bulgaria, 2017
5. R. Patyńska, Rock bursts and tectonics of the Upper Silesian Coal Basin. XXVII Zimowa Szkoła Mechaniki Górotworu, Wyd. Katedry Geomechaniki Budownictwa i Geotechniki Tom II, Kraków. p. 741-753 (2004)
6. R. Patyńska, Mining and geological conditions of exploitation of seams with rock burst hazard between 1987 and 2007. Wydawnictwo IGSMiE PAN, Gospodarka Surowcami Mineralnymi, Kwartalnik, Vol. 24- Book 2/3, Kraków, p. 227-243 (2008)
7. Z. Burtan, The importance of associated natural hazards in the Upper Silesian Coal Basin. Przegląd Górniczy, Vol. 11, p. 21-29 (2016)
8. R. Patyńska, K. Stec, Regional Rockburst Indicator for structural units of Upper Silesian Coal Basin. Studia Geotechnica et Mechanica, Vol. 39 (2017). No. 2, Wrocław 2017, p. 1-14 (2017)
9. K. Stec, Characteristics of Seismic Activity of the Upper Silesian Coal Basin in Poland. Geophysical Journal International, Blackwell Publishing Ltd, V 168. p. 213-219 (2007)
10. K. Stec, Seismic activity of Upper Silesian Coal Basin and its relationship with the geological properties of rock mass and parameters of mining operation. [in]: Stateczność Górotworu i obudowy przy łącznym obciążeniu statycznym i dynamicznym" (ed. A. Kidybiński), Wyd. Central Mining Institute, Katowice, p. 8-48 (2009)
11. J. Dubiński, W. Konopko, Rockbursts - assessment - forecast – fighting. Wydawnictwo GIG, Katowice (2000)
12. W. Konopko, R. Patyńska, Conditions of occurrence of rock bursts in coal mines. Przegląd Górniczy, Vol. 1, p. 12–17 (2008)
13. W. Konopko et al., Work safety in coal mines. Wydawnictwo GIG, Vol 2, Katowice, p.169-230 (2013) [12]

Geotechnical aspects of locating the Łagiewnicka Route on the area of the "White Seas" in Kraków

Jakub Zięba[1*], *Mirosława* Bazarnik[1,]

[1]Cracow University of Technology, Faculty of Civil Engineering, Krakow, Poland

Abstract. The article presents the results of field studies and laboratory tests of the subsoil of the section of the „Łagiewnicka" route located on the area of so-called "White Seas" [1] in Kraków-Łagiewniki. The analysed samples come from the layer of anthropogenic soils being the result of the operation of the Solvay Sodium Plant in the 20th century. The results of the geotechnical and land-surveying studies served as the basis for developing a numerical model to determine the pressure on the walls of the planned tunnel and the horizontal thrust in the planned excavation. Besides, the study and test results were used to perform an analysis of the excavation's stability and to check the impact of the sheet pile on land deformations near the John Paul II Centre in Kraków.

Keywords: Łagiewnicka Route, slope stability, MCC model

1 Introduction

The construction of so-called "Łagiewnicka Route" means the construction of a road section totalling 3.5 km in length. The beginning of the route is located at the intersection of Grota Roweckiego Street with the planned „Pychowicka Route" (now Norymberska Street) whilst the end of the route is located at the intersection of Jerzego Turowicza and Wincentego Witosa Streets. The Łagiewnicka Route is part of so-called "III Krakow Ringroad" and the construction of the whole section is planned for 2018-2020. The complex geotechnical conditions along the planned section and the terrain layout characteristics of this region, require the construction of 7 tunnel facilities, including 4 two-level intersections.

Due to the planned construction of a road tunnel, located in part on the anthropogenic soils of the "White Seas", a study and analysis were performed of the possibilities to make an excavation and the tunnel itself and of the impact of the works on the surroundings. Due to low strength parameters [2,3] of the limestone sediments in the sedimentation tanks, throughout this length, the tunnel will be located on quaternary or tertiary layers

* Corresponding author: jzieba@pk.edu.pl

immediately below the industrial sediments. The detailed route of the Łagiewnicka Route is presented in Fig. No. 1.

Since the tunnel construction work was done using the cut-and-cover method, at the section of the Łagiewnicka Route concerned, an attempt was made to assess the impact of the earthworks on the behaviour of the anthropogenic soils and the negative impact of the excavation on the John Paul II Centre in Kraków. The determination of the geotechnical parameters of the land enabled the construction of a numerical model which made it possible to determine the impact of the excavation on the behaviour of the land in its neighbourhood. The analyses served the purpose of presenting the results of the land deformation parameters on the area of the "White Seas".

2 Location and geological engineering conditions

The places where the samples were collected are marked on the satellite image with the Łagiewnicka Route marked thereon (Fig. 1). Besides, archival studies done by the Soil – Structure Interaction Division, Cracow University of Technology, for the area were also used.

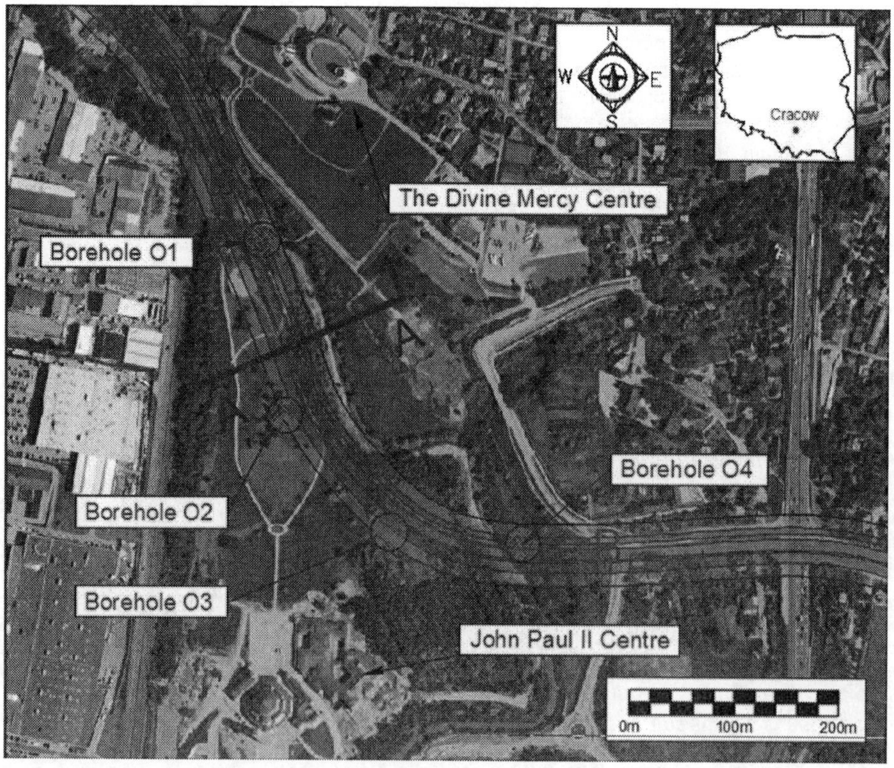

Fig. 1. Location of the "Łagiewnicka Route" course in the area of Solvay Sodium Plant landfill, also indicating the location of boreholes O1, O2, O3 and O4 [4,5]

The geological structure of the terrain, together with a description of the formation of the production sedimentation tanks of the Solvay Sodium Plant in recent years, was discussed in several papers [1,2,3,6]. In the subsoil layer, there are quaternary soils (sand and gravel alternating with cohesive soils, mainly dusts) which are laid out directly on the tertiary formations represented by Miocene silt. Above the indigenous soil, there is waste generated during the former production of the Solvay Sodium Plant in Kraków. The production generated waste which was stored in whole, in the form of suspended matter, in sedimentation tanks. The tanks covered a total area of more than 30 ha. The height of the stored waste was up to 20m.

Based on the field geotechnical and land-surveying studies which were done and an analysis of data in the writings [2,3,4], the following cross-sections were made: transverse A-A and longitudinal B-B for the section of the tunnel located on the area of the "White Seas" in Łagiewniki. The cross-sections took account of the geometry of the tunnel. Besides, it was proposed to extend the tunnel walls to locate them on the load-bearing layers.

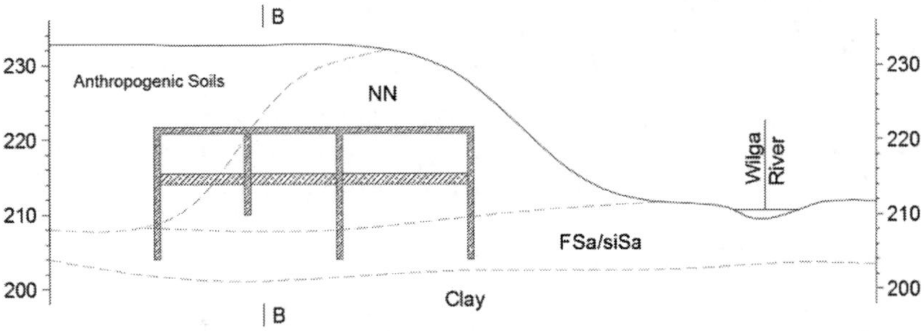

Fig. 2. Transversal cross-section A-A

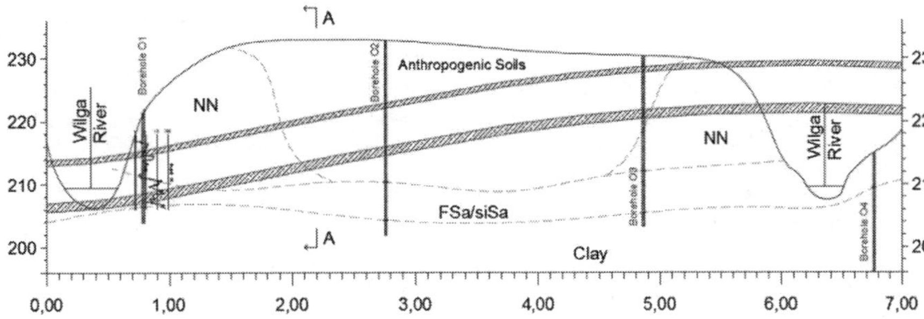

Fig. 3. Longitudinal cross-section B-B

The cross-sections and the specified parameters of the subsoil served, ultimately, to develop a numerical model with which a numerical analysis was done concerning the next stages of tunnel construction. The analysis takes account of the impact of the excavation, at the construction stage, on deformation of the subsoil in the neighbourhood of the John Paul II Centre in Kraków.

3 Material and Methodology

The soil for laboratory tests was collected from two test boreholes O2 and O3. In both of them, below the gravel layer, anthropogenic soils were found, mainly in a semi-compact and plastic condition. No ground water table was found in the boreholes. The water content of the soil in the profile varied.

After it had been transported to the laboratory, the soil was cold stored at a temperature of 5°C until the samples were formed and the tests were done. The tests were done on samples with natural water content. The test range was selected on the basis of the parameters necessary for developing a numerical model of digging the Łagiewnicka Route tunnel.

4 Results and analysis

The tests were done at the Laboratory Soil – Structure Interaction Division, Cracow University of Technology. The averaged results of the physical parameters which were used in the development of the numerical model are presented below.

4.1 Physical properties of the soil

The results of the physical parameters, i.e. bulk density, particle density, water content, porosity and the degree of saturation are presented in Table No. 1.

Table 1. Physical parameters of the soil on the area of the "White Seas" in the natural condition.

Sample	Bulk density [g/cm^3]	Water content [%]	Particle density [g/cm^3]	Porosity [-]	Degree of saturation [-]
Anthropogenic Soil	1,45	75	2,64	0,65	0,79
FSa/siSa	1,95	21	2,65	0,39	0,64
Clay	2,05	34	2,68	0,43	1,21

4.2 Mechanical properties of the anthropogenic soil

4.2.1 Shear strength

The tests of shear strength setting the critical angle of repose and coherence, were done in the shearing test apparatus. The compression stresses were applied of 50, 100, 150 and 200 kPa, accordingly. The shear results are presented in Fig. No. 4.

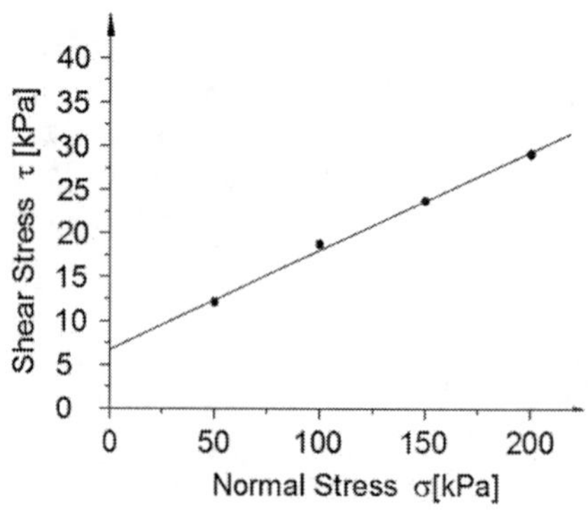

Fig. 4. Results of the shearing test.

Ultimately, based on tests for anthropological tests, the critical angle of repose was about 5 degree and the coherence about 8 kPa. The tests were done for 5 soil samples totalling 20 tests from each borehole.

4.2.2 Consolidation

Due to the use of the Modified Cam-Clay numerical model in the Midas GTS NX software, consolidation tests were done with the use of a oedometer. The test results for the representative sample with the specified path are presented in Fig. No. 5.

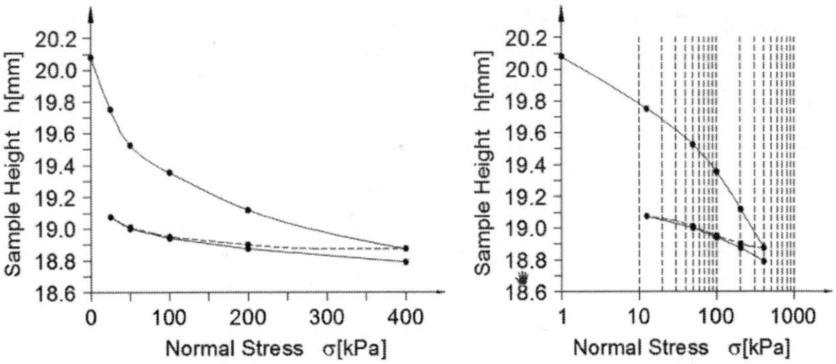

Fig. 5. Results of oedometric tests.

The figures for the primary deformation modules for the samples were 4÷9 MPa. The secondary deformation module was between 12 and 18 MPa.

4.2.3 Dynamic parameters of the soil

The setting of the dynamic parameters of the soil base from the area of the "White Seas" was the subject of the studies done in 2017. The results are presented in paper [3]. The dynamic module figures of the base in its natural condition, i.e. the module of non-dilatational strain G was set at 12-20 MPa. The module figure changed together with the depth in the test borehole. In a comparison of the results of the transversal and longitudinal wave, the Poisson ratio was set for soil with natural water content which is 0.35÷0.39 [3]. The parameters are presented in Table No. 2.

5 Numerical Analysis

The numerical analysis was done with the assumption that the tunnel would be constructed in the open excavation technology. Such technology for the construction of tunnels all alongside the "Łagiewnicka Route" was adopted at the stage of identifying the contractor for the route [4]. The considerable overburden in the area of the "White Seas" of even 12m above the upper tunnel plate caused a problem in the protection of the excavation walls along the section concerned. Due to the close proximity of the John Paul II Centre, an analysis was done of protecting the tunnel excavation, in three alternatives. Also, an analysis of stability of the alternatives was done to see whether the technology of doing the work would not have a negative effect on the location of the John Paul II Centre in Kraków.

The Modified Cam-Clay model was applied, which was based on the theory of resilience and plasticity theory with enhancement [7]. Non-linear behaviour is represented by an increase in the bulk modulus during the application of pressure in the material. Material plastification is applied in the material together with the destruction surface whose changes depend on the hardening or the softening of the material. Figure 6 presents a diagram of the relationship between the changes in the sample volume conditioned by the pressure in the MCC model.

https://doi.org/10.1051/e3sconf/20183603006

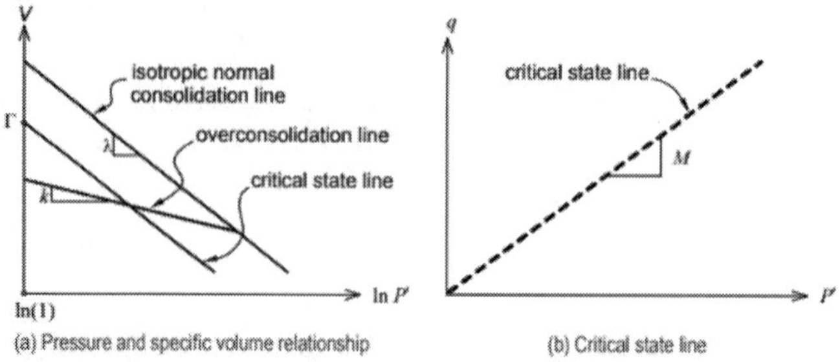

Fig. 6 Relationship between sample volume and pressure in the MCC model [8]

Detailed information on the assumptions in the MCC model for the soil can be found in the records [8,9,10]. Figure No. 7 presents the model with a simulation of the construction work staging. The symmetrical part of the structure was taken account of. The analysed area was implemented in the Midas GTS NX programme based on the finite element method.

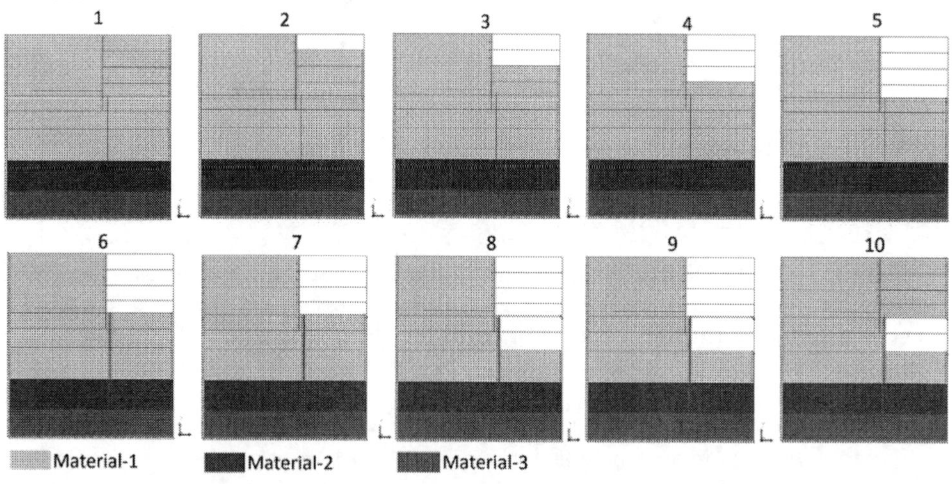

Fig. 7. Numerical model for the area, with staging No. 1-10

The geometry of the model covered an area of 32.5m in width and 35m in depth. The model was discretised with quadrangular finite elements of 0.5m. Non-displaceable boundary conditions were applied at the model's base. The vertical edges were blocked horizontally. The geotechnical layer with the thickness of 24 m, made of anthropological soil, was modelled with the MCC model. At the depth of between 24m and 30m, a clay layer was implemented in the MC model. The silt layers below 30m, were modelled with the MC model. Larssen type walls modelled with beam elements were selected as the encasing of the first excavation. The thrust depth was set at 14m below the terrain level. Cavity walls made of C30/37 concrete with the thickness of 0.8m were taken as the encasing of the other tunnel. The upper and lower plates with the thickness of 0.6m were modelled with beam elements. The contact was included between beam elements and the geotechnical layers. The following calculation alternatives were applied:

- Alternative I: checking the stability of the excavation to the level of the tunnel's upper plate
- Alternative II: designing the wall protecting the excavation in the form of a Larssen 755 wall.

- Alternative III: application of a vertical encasing of the planned tunnel made of a concrete cavity wall and the making of the excavation together with the construction of the tunnel's upper and lower plates. Covering the tunnel with anthropogenic soil.

The parameters of the materials applied in the numerical simulations are presented in Table No. 2.

Table 2. Materials parameters used in numerical simulations.

Material:	Material 1	Material 2	Material 3	Steel Larssen 755	Concrete C30/37
Model:	MCC	MC	MC	Elastic	Elastic
E [MPa]	9.27	80	80	210000	33000
ν [-]	0.37	0.30	0.30	0.30	0.20
γ[kN/m^3]	14.22	21.58	21.09	78.50	24.00
φ [°]	5	20	26	-	-
c [kPa]	8	35	65	-	-
K_0 [-]	0.58	0.43	0.43	1.0	1.0
λ	0.002	-	-	-	-
κ	0.001	-	-	-	-
M	0.179	-	-	-	

Numerical simulations for alternative I demonstrated that an unprotected excavation with appropriate inclination of the slopes is only possible at the inclination of 1:4.5. Due to a low efficiency of this solution, it is recommended to protect the excavation in accordance with alternatives II and III. Figure 8 presents the results of the initial inclination of the slopes of 1:1 (Fig. 8a), and the inclination of 1:4.5 stabilising the excavation slope (Fig. 8b).

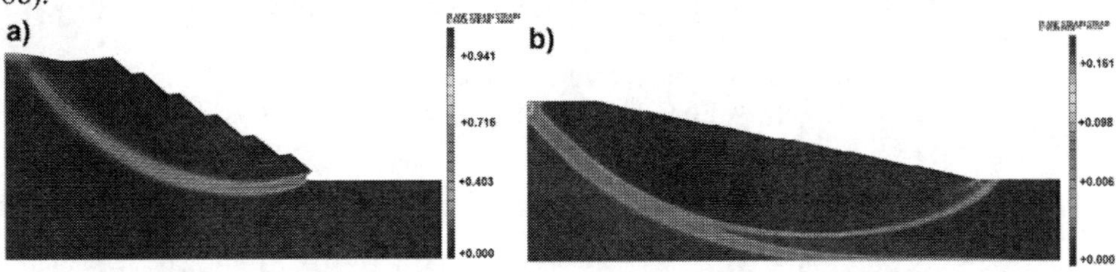

Fig. 8. a) Slip surface for the inclination alternative 1:1 and FoS=0.54 b) Slip surface for the inclination of 1:4.5 and FoS=1.05

The numerical simulations for alternatives II and III were based on the following stages (Fig.7):

- Stage 1: thrusting a sheet pile to the depth of 14m below the terrain level,
- Stages 2,3,4,5: making of an excavation to the depth of 3m, 6m, 9m, 11.5m below terrain level, with the protection of the wall with ground anchors,
- Stage 6: making of a cavity wall to the depth of 18.5 m from the excavation level,
- Stage 7: construction of the upper plate,
- Stage 8: extracting the soil under the plate,
- Stage 9: construction of the tunnel's lower plate,
- Stage 10: covering the tunnel with anthropogenic soil.

With the application of appropriate support of the cavity wall, the analysis demonstrated the overall effectiveness of the alternative which was adopted, as a result of which the land in the neighbourhood of the CJPII will not be subject to excessive deformations.

6 Conclusions

The paper presents the results of field studies and laboratory studies and tests of the parameters of soils from the production waste heap on the area of the defunct Solvay Sodium Plant in Kraków. The results of laboratory tests together with the results of field studies done to identify the base for the John Paul II Centre (CJPII) project in Łagiewniki, in Kraków served the purpose of developing a numerical model and analysing the stage of construction work on the Łagiewnicki Route tunnel on land deformation in the immediate neighbourhood of the John Paul II Centre.

The area concerned is characterised by low geotechnical parameters. This is confirmed by the Soil – Structure Interaction Division own studies which were part of preparation of the execution documentation for the Łagiewnicka Route tunnel on the area of the "White Seas" [4]. Due to low soil strength parameters, it can finally be concluded that the land cannot be used as the soil bearing substrate to locate the future road project there.

Besides, the results of the soil numerical model, which take account of the base parameters specified on the basis of the studies and tests, point to the need to protect the excavation to counteract the negative impacts on the John Paul II Centre.

6 References

1. I. Krzak, Zagospodarowanie terenów poprzemysłowych Krakowskich Zakładów Sodowych „Solvay", Problemy Ekologii Krajobrazu, Tom 17, s.283÷287 (2005)
2. B. Czado, Analiza nośności pali fundamentowych na podstawie polowych badań gruntów sondą statyczną, Rozprawa doktorska, Biblioteka Politechniki Krakowskiej (2014)
3. E. Pilecka, J. Zięba "Preliminary studies of the dynamic stiffness modules of soil samples from the Solvay Sodium Plant waste landfill in Krakow", E3S Web of Conferences (2017)
4. http://trasalagiewnicka.krakow.pl/
5. http://www.geoportal.gov.pl//
6. W. Sroczyński, Perspektywy zagospodarowania tzw. "białych Mórz" na terenach po byłych krakowskich Zakładach sodowych "Solvay" Komisja Krajobrazu Kulturowego PTG s.423÷430 (2008).
7. A. M., Britto, M. J. Gunn, Critical state soil mechanics via finite elements, Ellishorwood Limited (1987)
8. Midas GTS NX, Manual specification. Materials, (2016)
9. T. Godlewski, T. Szczepański, Metody określania sztywności gruntów w badaniach geotechnicznych, Instytut Techniki Budowlanej, Warszawa (2015)
10. R. I., Borja, Cam-clay plasticity, Part II: Implicit integration of constitutive equation based on a nonlinear elastic stress predictor, COMPUT METHOD APPL M, Vol. 88, Issue 2, pp. 225-240 (1991)

AUTHOR INDEX